Series in Real Analysis – Vol. 11

GENERALIZED ORDINARY DIFFERENTIAL EQUATIONS

Not Absolutely Continuous Solutions

SERIES IN REAL ANALYSIS

Series in Real Analysis – Vol. 11

GENERALIZED ORDINARY DIFFERENTIAL EQUATIONS

Not Absolutely Continuous Solutions

Jaroslav Kurzweil

Academy of Sciences, Czech Republic

 **World Scientific**

NEW JERSEY · LONDON · SINGAPORE · BEIJING · SHANGHAI · HONG KONG · TAIPEI · CHENNAI

Published by

World Scientific Publishing Co. Pte. Ltd.

5 Toh Tuck Link, Singapore 596224

USA office: 27 Warren Street, Suite 401-402, Hackensack, NJ 07601

UK office: 57 Shelton Street, Covent Garden, London WC2H 9HE

British Library Cataloguing-in-Publication Data
A catalogue record for this book is available from the British Library.

Series in Real Analysis — Vol. 11
GENERALIZED ORDINARY DIFFERENTIAL EQUATIONS
Not Absolutely Continuous Solutions

Copyright © 2012 by World Scientific Publishing Co. Pte. Ltd.

For photocopying of material in this volume, please pay a copying fee through the Copyright Clearance Center, Inc., 222 Rosewood Drive, Danvers, MA 01923, USA. In this case permission to photocopy is not required from the publisher.

ISBN-13 978-981-4324-02-1
ISBN-10 981-4324-02-7

Printed in Singapore.

Preface

The Kurzweil-Henstock integral is a nonabsolutely convergent integral. The aim of this treatise is the exploitation of this property in some convergence problems in ordinary differential equations and in some situations where solutions of infinite variation can occur. This leads to generalized differential equations the theory of which is presented in Chapters 6–18, 22–24, 27. Since the motion of Kapitza's pendulum was a significant motivation for the theory, the equation of motion of Kapitza's pendulum is exposed in Chapter 2 and various estimates for equations of a similar type are obtained by elementary methods in Chapters 3 and 4.

The concept of the Kurzweil-Henstock integral and of the generalized differential equation goes back to 1957. The preparation of this book started in 2004 after W. N. Everitt had encouraged me to build a solid theory of the generalized differential equations. The results in this direction were reported and discussed regularly in the Seminar on Differential Equations and Integration Theory in the Institute of Mathematics of the Academy of Sciences of the Czech Republic in Prague.

I wish to thank the participants of the seminar for their contributions and comments, in particular to J. Jarník, B. Maslowski, I. Vrkoč, M. Tvrdý and the late Š. Schwabik. The last two of them in addition read and commented the manuscript. E. Ritterová was very helpful by typing several versions of the manuscript.

The research connected with the preparation of this book was supported by the grant No. IAA100190702 of the Grant Agency of the Acad. Sci. of the Czech Republic and by the Academy of Sciences of the Czech Republic, Institutional Research Plan No. AV0Z10190503.

Prague, May 2011   Jaroslav Kurzweil

Contents

Chapter 1

Introduction

A solution of a classical differential equation

$$\dot{x} = f(x,t) \tag{1.1}$$

is a function u such that its derivative $\dot{u}$ is at every τ equal to $f(u(\tau),\tau)$, i.e. in a neighborhood of τ the linear function

$$t \to u(\tau) + f(u(\tau),\tau)\,(t-\tau)$$

is a good approximation of u. Usually f and u are $\mathbb{R}^n$-valued functions. Given $u(a) = y$ the value $u(T)$ is approximately equal to

$$y + \sum_{i=1}^{k} f(u(\tau_i),\tau_i)\,(t_i - t_{i-1})$$

for $T > a$. Here

$$a = t_0 < t < \cdots < t_k = T, \tag{1.2}$$

$\tau_i \in [t_{i-1}, t_i]$, τ_i being called the tag of the interval $[t_{i-1}, t_i]$ and the partition of $[a,b]$ into intervals $[t_{i-1}, t_i]$ is sufficiently fine. Moreover, u is a solution of the Volterra integral equation

$$u(T) = y + \int_a^T f(u(t),t)\,\mathrm{d}t \tag{1.3}$$

and vice versa, every solution of (1.3) is a solution of (1.1) and fulfils $u(a) = y$.

A generalized ordinary differential equation (GODE)

$$\frac{\mathrm{d}}{\mathrm{d}t}x = \mathrm{D}_t F(x,\tau,t) \tag{1.4}$$

depends on a function F of three variables and its solution is a function u such that the function

$$t \to u(\tau) + F(u(\tau), \tau, t) - F(u(\tau), \tau, \tau)$$

is a good approximation of u in a neighborhood of any τ. The value $u(T)$ is approximately equal to

$$u(a) + \sum_{i=1}^{k} [F(u(\tau_i), \tau_i, t_i) - F(u(\tau_i), \tau_i, t_{i-1})]. \qquad (1.5)$$

In fact, the sum in (1.5) can be viewed as an approximation of an integral which is denoted by

$$\int_a^T D_t F(u(\tau), \tau, t). \qquad (1.6)$$

The quality of approximation depends on the interpretation of the concept that the partition of $[a, T]$ is fine.

By definition, u is a solution of (1.4) if it fulfils

$$u(T) = u(a) + \int_a^T D_t F(u(\tau), \tau, t) \qquad (1.7)$$

for $T > a$, which is a Volterra-type integral equation.

The concept of a fine partition of $[a, b]$ admits various interpretations and two of them are crucial in this treatise.

Let $\xi > 0$, $[a, T] \subset \mathbb{R}$. A set $\{([t_{i-1}, t_i], \tau_i) ; i = 1, 2, \ldots, k\}$ is a ξ-fine partition of $[a, T]$ if (1.2) holds, if $t_i - t_{i-1} \leq \xi$ and $\tau_i \in [t_{i-1}, t_i]$ for $i = 1, 2, \ldots, k$.

Let δ be a positive function on $[a, T]$, i.e. $\delta : [a, T] \to \mathbb{R}^+$. A set $\{([t_{i-1}, t_i], \tau_i) ; i = 1, 2, \ldots, k\}$ is a δ-fine partition of $[a, T]$ if (1.2) holds and if

$$\tau_i - \delta(\tau_i) \leq t_{i-1} \leq \tau_i \leq t_i \leq \tau_i + \delta(\tau_i) \quad \text{for } i = 1, 2, \ldots, k.$$

These two concepts of a fine partition of $[a, T]$ are a basis for two concepts of a solution of (1.4).

Let us proceed to the concept of a solution u of (1.4) without making precise the integral in (1.6). Let X be a Banach space, $\|x\|$ denoting the norm of x for $x \in X$. Assume that $[a, b] \subset \mathbb{R}$, $u : [a, b] \to X$ and that there exists $\xi_0 > 0$ such that $F(u(\tau), \tau, t)$ is defined for $\tau, t \in [a, b]$,

$|t - \tau| \leq \xi_0$. u is called an SR-solution (Strong Riemann solution) of (1.4) if for every $\varepsilon > 0$ there exists $\xi > 0$ such that

$$\sum_{i=1}^{k} \|u(t_i) - u(t_{i-1}) - F(u(\tau_i), \tau_i, t_i) + F(u(\tau_i), \tau_i, t_{i-1})\| \leq \varepsilon$$

for every ξ-fine partition $\{([t_{i-1}, t_i], \tau_i); i = 1, 1, \ldots, k\}$ of $[a, b]$.

Assume that $[a, b] \subset \mathbb{R}$, $u: [a, b] \to X$ and that there exists $\delta_0 : [a, b] \to \mathbb{R}^+$ such that $F(u(\tau), \tau, t)$ is defined for $\tau, t \in [a, b]$, $|t - \tau| \leq \delta_0(\tau)$. u is called an SKH-solution (Strong Kurzweil-Henstock solution) of (1.4) if for every $\varepsilon > 0$ there exists $\delta : [a, b] \to \mathbb{R}^+$ such that

$$\sum_{i=1}^{k} \|u(t_i) - u(t_{i-1}) - F(u(\tau_i), \tau_i, t_i) + F(u(\tau_i), \tau_i, t_{i-1})\| \leq \varepsilon$$

for every δ-fine partition $\{([t_{i-1}, t_i], \tau_i) ; i = 1, 2, \ldots, k\}$ of $[a, b]$. This definition is correct since for every $\delta : [a, b] \to \mathbb{R}^+$ there exists a δ-fine partition of $[a, b]$. Let $\xi_0 > 0$, $0 < \xi < \xi_0$, and let $\delta(\tau) \leq \xi$ for $\tau \in [a, b]$. Then every δ-fine partition of $[a, b]$ is a ξ-fine partition of $[a, b]$. Therefore every SR-solution of (1.4) is an SKH-solution of (1.4) but not vice versa.

The origin of the concept of a GODE goes back to the averaging principle by which the solutions of

$$\dot{x} = h_0(x) + \sum_{j=1}^{k} h_j(x) \cos(\sigma_j t/\varepsilon + \eta_j), \quad (\sigma_j \neq 0) \tag{1.8}$$

tend for $\varepsilon \to 0$ to the solutions of

$$\dot{x} = h_0(x) \tag{1.9}$$

which is called the averaged equation since

$$h_0(x) = \lim_{T \to \infty} \frac{1}{T} \int_0^T \left[h_0(x) + \sum_{j=1}^{k} h_j(x) \cos(\sigma_j t/\varepsilon + \eta_j) \right] dt .$$

The right hand side of (1.8) does not converge pointwise for $\varepsilon \to 0$ which gave an impulse to tying together the convergence of solutions with various types of convergence of the right hand sides of the differential equations.

The book consists of five parts. In the first part (Chapters 2–4) Kapitza's pendulum is briefly treated. Its equation is transformed to an equation of the form (1.8). Let $\lambda_i : \mathbb{R} \to \mathbb{R}$, $i = 1, 2, \ldots, k$,

$$\lambda_j(t+1) = \lambda_j(t), \quad \int_0^1 \lambda_j(s)\, ds = 0, \quad h_j : \mathbb{R}^n \to \mathbb{R}^n.$$

Let $u : [a,b] \to \mathbb{R}^n$ be a solution of

$$\dot{x} = h_0(x) + \sum_{j=1}^n h_j(x)\,\lambda_j(t/\varepsilon), \tag{1.10}$$

and let $w : [a,b] \to \mathbb{R}^n$ be a solution of (1.9), $u(a) = w(a)$. Then

$$\|u(t) - w(t)\| \le \varkappa \varepsilon \quad \text{for } t \in [a,b]$$

for some $\varkappa > 0$ and ε sufficiently small, i.e. w is an approximation of u of order ε. Moreover, an approximation $\widetilde{w}$ of u of order ε^2 can be defined by the formula

$$\widetilde{w}(t) = w(t) + \sum_{j=1}^k h_j(w(t))\,\Lambda_j(t/\varepsilon)$$

where $\Lambda_j : \mathbb{R} \to \mathbb{R}$ is the primitive of λ_j such that $\int_0^1 \Lambda_j(s)\, ds = 0$. (Observe that $\widetilde{w}(a) = w(a) + \varepsilon \sum_{j=1}^k h_j(w(a))\,\Lambda_j(a/\varepsilon)$.)

The second part (Chapters 5–13) contains a theory of GODEs in the form

$$\frac{d}{dt} x = D_t G(x, \tau, t) \tag{1.11}$$

where G fulfils conditions (8.2)–(8.7). In simplified form conditions (8.3), (8.4) may be written as

$$\|G(x,\tau,t) - G(x,\tau,s)\| \le \varkappa |t-s|^\alpha, \tag{1.12}$$

$$\|G(x,\tau,t) - G(x,\tau,s) - G(y,\tau,t) + G(y,\tau,s)\| \le \varkappa \|x-y\|\,|t-s|^\beta, \tag{1.13}$$

where $\varkappa > 0$, $0 < \alpha \le \beta$, $\alpha + \beta > 1$. Conditions (8.5)–(8.7) are of a similar type. A local existence theorem and a continuous dependence theorem are valid for the equation of the type (1.11). By the continuous dependence theorem there exists a function $\Omega : \mathbb{R} \to \mathbb{R}$ such that $\Omega(\eta) \to 0$ for $\eta \to 0$ and the following situation takes place: Let G^* fulfil (8.1)–(8.7), let $v : [a, b] \to X$ be a solution of (1.11) and let $v^* : [a, b] \to X$ be a solution of

$$\frac{d}{dt} x = D_t G^*(x, \tau, t), \qquad (1.14)$$

$v(a) = v^*(a)$. If $G\|(x, \tau, t) - G^*(x, \tau, t)\| \leq \eta$ (for (x, τ, t) from a suitable set) then $\|v(t) - v^*(t)\| \leq \Omega(\eta)$ for $t \in [a, b]$. A uniqueness result is postponed to Chapter 16 since it is proved for the concept of SKH-solutions which is wider than the concept of SR-solutions.

A solution $u : [a, b] \to X$ of (1.11) fulfils $\|u(t) - u(s)\| \leq \varkappa_1 |t - s|^\alpha$, where $\varkappa_1 > 0$, $t, s \in [a, b]$, but need not have finite variation on any interval. By definition u is a solution of the corresponding integral equation of Volterra type.

The main object of the third part (Chapters 14–18) is the concept of an SKH-solution of the GODE (1.11), no specific assumptions on G being imposed. In particular:

(i) If $a < c < b$, $u : [a, b] \to X$ and if u is an SKH-solution of (1.11) on $[a, c]$ and on $[c, b]$ then it is an SKH-solution on $[a, b]$ (which need not hold for SR-solutions in general).

(ii) If $u : [a, b] \to X$ is an SKH-solution of (1.11) on $[c, b]$ for every c such that $a < c < b$ and if $\big[u(c) - u(a) - G(u(a), a, c) - G(u(a), a, a) \big] \to 0$ for $c \to a$, then u is an SKH-solution of (1.11) on $[a, b]$.

Let g depend on the variables x, t and $u : [a, b] \to X$. Then u is called an SKH-solution of the differential equation in the classical form

$$\dot{x} = g(x, t) \qquad (1.15)$$

if

$$u(T) - u(S) = (\text{SKH}) \int_S^T D_t[g(u(\tau), \tau) t] = (\text{SKH}) \int_S^T g(u(t), t) \, dt.$$

Conditions on g, G are found such that u is a solution of (1.15) if and only if it is a solution of (1.11).

The following result is an illustration of the above concepts:

Let $h : X \times (\mathbb{R} \setminus \{0\}) \to X$, $H : X \times (\mathbb{R} \setminus \{0\}) \to X$, the main assumption being that h and H are bounded and continuous,

$$\frac{\partial}{\partial t} H(x,t) = h(x,t) \quad \text{and} \quad 1 \le \alpha < 1 + \tfrac{1}{2}\beta.$$

Put

$$g(x,t) = \begin{cases} t^{-\alpha} h(x, t^{-\beta}) & \text{if } t > 0, \\ 0 & \text{if } t = 0, \\ |t|^{-\alpha} h(x, -|t|^{-\beta}) & \text{if } t < 0. \end{cases}$$

Then for every $y \in X$ there exists a unique SKH-solution $u : [a,b] \to X$ of (1.15), $u(a) = y$. u need not be absolutely continuous on any neighbourhood of 0.

Let $\Phi : [a,b] \to \mathbb{R}$ be nondecreasing, $\Phi(\tau) - \lim_{t \to \tau, t < \tau} \Phi(t) < 1$ for $a < \tau \le b$. The fourth part (Chapters 19–24) contains a theory of the GODE

$$\frac{\mathrm{d}}{\mathrm{d}t} x = F(x, \tau, t), \tag{1.16}$$

where $F : X \times [a,b]^2 \to X$ fulfils

$$\| F(x, \prime, t) - F(x, \prime, s) \| \le (1 + \|x\|) \left| \Phi(t) - \Phi(s) \right|$$

and three conditions of a similar type. For every $y \in X$ there exists a unique SKH-solution $u : [a,b] \to X$ of (1.16), $u(a) = y$. This solution is a function of bounded variation and depends continuously on y and F.

Put

$$F^\circ(x, T) = (\text{SKH}) \int_a^T D_t F(x, \tau, t).$$

Then u is an SKH-solution of (1.16) if and only if it is an SKH-solution of

$$\frac{\mathrm{d}}{\mathrm{d}t} x = F^\circ(x, t). \tag{1.17}$$

In other words: a process which is described by equation (1.16) is described by equation (1.17) as well. Moreover, for a given F there may exist a sequence of functions F_i, $i \in \mathbb{N}$, which are continuous and fulfil similar conditions as F and if $u_i : [a,b] \to X$ is a solution of

$$\frac{\mathrm{d}}{\mathrm{d}t} x = D_t F_i(x, \tau, t), \quad u_i(a) = y,$$

then u_i is continuous, the sequence u_i is bounded and $u_i(t) \to u(t)$ almost everywhere. Observe that Chapter 19 contains only such results which follow directly from the definition of the SKH-solution and that rather general results on the existence of (SKH) $\int_a^b D_t U(\tau, t)$ and on integration-by-parts are presented in Chapters 20,21.

The main topic of the fifth part (Chapters 25–27) is the treatment of the GODEs

$$\frac{d}{dt} x = D_t G(x, \tau, t),$$ (1.18)

$$\frac{d}{dt} x = D_t G^\circ(x, \tau, t),$$ (1.19)

where $G \colon X \times [a,b]^2 \to X$ fulfils conditions which are similar to the conditions imposed on G in the GODE (8.8) and

$$G^\circ(x,t) = (\text{SR}) \int_a^t D_t G(x, \sigma, s).$$ (1.20)

In the simplified form G fulfils

$$\|G(x,\tau,t) - G(x,\tau,s)\| \leq \varkappa \left(1 + \|x\|\right) |t - s|^\alpha \quad \text{for } x \in X,\ \tau, t, s \in [a,b],$$

$$\|G(x,\tau,t) - G(x,\tau,s) - G(z,\tau,t) + G(z,\tau,s)\| \leq \varkappa \|x - z\| \, |t - s|^\beta$$
$$\text{for } x, z \in X,\ \tau, t, s \in [a,b],$$

where $\varkappa > 0$, $0 < \alpha \leq \beta$, $\alpha + \beta > 1$, and three similar conditions. The integral in (1.20) exists and G° fulfils analogical conditions as G. Moreover, u is a solution of (1.18) if and only if it is a solution of (1.19).

The main part of the book was written in the years 2007-2009. The results on the relation of the GODEs (1.15), (1.16) and on the relation of the GODEs (1.18), (1.19) were added to the text in the final phase of its preparation.

Chapter 2

Kapitza's pendulum
and a related problem

In [Kapitza (1951a)], [Kapitza (1951b)] the author studied the equation

$$\ddot{\Theta} = (g\,L^{-1} - L^{-1}\,(A\omega)\,\omega\,\sin\,\omega t)\,\sin\,\Theta \qquad (2.1)$$

which describes the motion of a pendulum the support of which is vibrating with a large frequency ω and a small amplitude A. Here g, L, ω, $A \in \mathbb{R}^+$, L is the length of the pendulum and g is the gravitational constant. Putting

$$\dot{\Theta} - L^{-1}\,(A\omega)\cos\,\omega t\,\sin\,\Theta = \Phi$$

the equation (2.1) is transformed to

$$\left.\begin{aligned}
\dot{\Theta} &= \Phi + L^{-1}\,A\omega\,\sin\,\Theta\,\cos\,\omega t,\\
\dot{\Phi} &= g\,L^{-1}\,\sin\,\Theta - \tfrac{1}{2}\,(L^{-1}A\omega)^2\,\sin\,\Theta\,\cos\,\Theta\\
&\quad -\tfrac{1}{2}\,(L^{-1}\,A\omega)^2\sin\,\Theta\,\cos\,\Theta\,\cos\,2\omega t\\
&\quad -L^{-1}\,A\omega\,\Phi\,\cos\,\Theta\,\cos\,\omega t.
\end{aligned}\right\} \qquad (2.2)$$

If ω is large and A is small (i.e. if $A\omega \leq$const.) then the solutions of (2.2) are close to the solutions of the averaged equation

$$\left.\begin{aligned}
\dot{\Theta} &= \Phi,\\
\dot{\Phi} &= g\,L^{-1}\,\sin\,\Theta - \tfrac{1}{2}\,(L^{-1}\,A\omega)^2\,\sin\,\Theta\,\cos\,\Theta.
\end{aligned}\right\} \qquad (2.3)$$

The relation of solutions of (2.2) and (2.3) can be studied as the relation of solutions of

$$\dot{x} = f(x) + h(x, t, t/\varepsilon), \quad \varepsilon > 0 \qquad (2.4)$$

and the solutions of

$$\dot{x} = f(x), \qquad (2.5)$$

the main assumption being the existence of a bounded function $H(x, \tau, t)$ such that

$$\frac{\partial}{\partial t} H(x, \tau, t) = h(x, \tau, t).$$

This problem will be tackled by elementary methods in Chapter 3. In Chapter 4 an equation of a more special form than (2.4) is treated but more precise results are obtained.

An insight into the problem of averaging is given by the equation

$$\dot{x} = x \, \varepsilon^{-\alpha} \, \cos(2\pi \, t/\varepsilon) + \varepsilon^{-\alpha} \, \sin(2\pi \, t/\varepsilon) \tag{2.6}$$

where $0 < \alpha \leq 1/2$. Let $y \in \mathbb{R}$ and denote by $u_{\alpha,\varepsilon} \colon [-1, 1] \to \mathbb{R}$ the solution of (2.6) fulfilling

$$x(0) = y. \tag{2.7}$$

It is well known that

$$u_{\alpha,\varepsilon}(t) = \exp\left(\frac{\sin(2\pi \, t/\varepsilon)}{2\pi \, \varepsilon^{\alpha-1}}\right) \left(y + \int_0^t \exp\left(-\frac{\sin(2\pi \tau/\varepsilon)}{2\pi \, \varepsilon^{\alpha-1}}\right) \frac{\sin(2\pi \tau/\varepsilon)}{\varepsilon^{\alpha}} \, \mathrm{d}\tau\right)$$

and it can be deduced that

$$\left| u_{\alpha,\varepsilon}(t) - y + \frac{\varepsilon^{1-2\alpha} \, t}{4\pi} \right| \leq \text{const.} \, \varepsilon^{1-\alpha}, \tag{2.8}$$

const. being independent of α, ε, t. Hence, for $\varepsilon \to 0$,

$$u_{\alpha,\varepsilon}(t) \to \begin{cases} y & \text{if } \alpha < \frac{1}{2}, \\ y - \dfrac{t}{4\pi} & \text{if } \alpha = \frac{1}{2}. \end{cases}$$

In other words, $u_{\alpha,\varepsilon}$ tends to a solution of $\dot{x} = 0$ if $\alpha < \frac{1}{2}$ and $\varepsilon \to 0$ while $u_{\alpha,\varepsilon}$ tends to a solution of $\dot{x} = -1/(4\pi)$ if $\alpha = \frac{1}{2}$ and $\varepsilon \to 0$. If $\alpha = 0$ then (cf. (2.8))

$$\|u_{1,\varepsilon}(t) - y\| \leq \text{const.} \, \varepsilon.$$

2.1. Remark. In [Lojasiewicz (1955)] the author treated an equation the particular case of which is (2.1) and obtained the corresponding convergence results. J. Jarník approached an analogous problem by means of the Kurzweil-Henstock integral and related ideas, see [Jarník (1965)]. A different method was used in [Mitropolskij (1971)].

Chapter 3

Elementary methods: averaging

The aim of this chapter is to estimate the difference of the solutions u_ε of

$$\dot{x} = f(x) + h(x, t/\varepsilon) \tag{3.1}$$

and the solutions w of the "averaged" differential equation

$$\dot{x} = f(x), \tag{3.2}$$

the main assumption being the existence of a *bounded* H such that

$$\frac{\partial}{\partial t} H(x, t) = h(x, t). \tag{3.3}$$

Moreover, it is assumed that there exist $\mu, \nu \in \mathbb{R}^+$ such that

$$f \text{ and its differential } Df \text{ are bounded by } \mu \tag{3.4}$$

and

$$h, H \text{ and some of their derivatives are bounded by } \nu. \tag{3.5}$$

Let $[a, b] \subset \mathbb{R}$. It will be proved in this chapter that there exist $\varkappa_1, \varkappa_2 \in \mathbb{R}$, depending only on μ and $b - a$, such that the inequality

$$\|u_\varepsilon(t) - w(t)\| \le \varepsilon \left(\nu \varkappa_1 + \nu^2 \varkappa_2 \right), \quad t \in [a, b] \tag{3.6}$$

holds for the solution u_ε of (3.1) and the solution w of (3.2) such that $u_\varepsilon(a) = w(a)$.

(3.6) is valid if ν is large provided $\varepsilon > 0$ is sufficiently small (cf. (3.7). Let $0 \le \alpha < \frac{1}{2}$ and let $u_{\alpha,\varepsilon}$ be a solution of

$$\dot{x} = f(x) + \varepsilon^{-\alpha} h(x, t/\varepsilon) \tag{3.7}$$

where f fulfils (3.4) and h, H fulfil (3.5). Then

$$\left. \begin{array}{l} \varepsilon^{-\alpha}h, \ \varepsilon^{-\alpha}H \quad \text{and some of their derivatives} \\ \qquad\qquad \text{are bounded by } \varepsilon^{-\alpha}\nu. \end{array} \right\} \quad (3.8)$$

(3.6) implies that

$$\|u_{\alpha,\varepsilon}(t) - w(t)\| \le \varepsilon^{1-\alpha}\,\nu\,\varkappa_2 + \varepsilon^{1-2\alpha}\,\nu^2\,\varkappa_3, \quad t\in[a,b] \qquad (3.9)$$

if $u_{\alpha,\varepsilon}(a) = w(a)$.

3.1. Notation. Let $\mathbb{R}$ denote the set of reals, $\mathbb{R}^+$ the set of positive reals, $\mathbb{R}_0^+$ the set of nonnegative reals, $\mathbb{N}$ the set of positive integers, $\mathbb{N}_0$ the set of nonnegative integers. Let

$$R,\ \mu,\ \nu,\ \varepsilon\in\mathbb{R}^+,\ n\in\mathbb{N}. \qquad (3.10)$$

Let $[a,b]$ denote the segment in $\mathbb{R}$ with endpoints a,b for $a,b\in\mathbb{R}$, $a<b$. The vector space $\mathbb{R}^n$ is supposed to be equipped with a norm, $\|x\|$ denoting the norm of x for $x\in\mathbb{R}^n$.

If $r\ge 0$ then

$$B(r) = \{x\in\mathbb{R}^n\,;\|x\|\le r\}$$

(i.e. $B(r)$ is the ball in $\mathbb{R}^n$ with center at 0 and radius r).

If $h:B(r)\to\mathbb{R}^n$ then $\mathrm{D}\,h(x)$ is the differential of h at x. Analogously, if $h:B(r)\times[a,b]\to\mathbb{R}^n$ then $\mathrm{D}_1\,h(x,t)$ is the differential of h with respect to x at (x,t) and $\mathrm{D}_2\,h(x,t)$ is the differential of h with respect to t at (x,t). $\mathrm{D}^2\,h(x)$ is the differential of the second order of h at x.

Let

$$f:B(R)\to\mathbb{R}^n,\quad h:B(R)\times\mathbb{R}\to\mathbb{R}^n,\quad H:B(R)\times\mathbb{R}\to\mathbb{R}^n. \qquad (3.11)$$

Assume that

$$f,\ \mathrm{D}\,f \quad \text{are continuous}, \qquad (3.12)$$

$$\|f(x)\| \le \mu,\ \|\mathrm{D}\,f(x)\| \le \mu \quad \text{for } x\in B(R), \qquad (3.13)$$

$$h,\ \mathrm{D}_1\,h,\ \mathrm{D}_1\,H,\ \mathrm{D}_2\,H \quad \text{are continuous}, \qquad (3.14)$$

$$\mathrm{D}_2\,H(x,t) - h(x,t) \quad \text{for } x\in B(R),\ t\in\mathbb{R} \qquad (3.15)$$

and

$$\|h(x,t)\|, \ \|\mathrm{D}_1 h(x,t)\|, \ \|H(x,t)\|, \ \|\mathrm{D}_1 H(x,t)\| \le \nu \\ \text{for } x \in B(R), \ t \in \mathbb{R}. \tag{3.16}$$

3.2. Remark. Conditions (3.13)–(3.16) may be fulfilled if

$$h(x,t) = \sum_{i=1}^{k} h_i(x)\,\lambda_i(t) \tag{3.17}$$

where $h_i : B(R) \to \mathbb{R}^n$, $\lambda_i : \mathbb{R} \to \mathbb{R}$ are continuous, if there exist $\omega_i \in \mathbb{R}^+$ such that $\lambda_i(t+\omega_i) = \lambda_i(t)$ for $t \in \mathbb{R}$ and $\displaystyle\int_0^{\omega_i} \lambda_i(t)\,\mathrm{d}t = 0$, $i = 1, 2, \ldots, k$, since the primitive Λ_i of λ_i is bounded and

$$H(x,t) = \sum_{i=1}^{k} h_i(x)\,\Lambda_i(t)\,.$$

3.3. Theorem. *Assume that $y \in B(R)$, that $u_\varepsilon : [a,b] \to B(R)$ is a solution of (3.1), $u(a) = y$ and that $w : [a,b] \to B(R)$ is a solution of (3.2), $w(a) = y$. Then*

$$\|u_\varepsilon(t) - w(t)\| \\ \le \varepsilon \left[(2+\mu)(b-a)\,e^{\mu\,(b-a)}\nu + (b-a)\,e^{\mu\,(b-a)}\,\nu^2 \right]. \tag{3.18}$$

Proof. Observe that (cf. (3.15))

$$\varepsilon \frac{\mathrm{d}}{\mathrm{d}t} H(u_\varepsilon(t), t/\varepsilon) = \varepsilon\,\mathrm{D}_1\,H(u_\varepsilon(t), t/\varepsilon)\,\dot{u}_\varepsilon(t) + h(u_\varepsilon(t), t/\varepsilon)\,.$$

Hence after substituting for $\dot{u}_\varepsilon(t)$ we obtain

$$\int_a^t h(u_\varepsilon(s), s/\varepsilon)\,\mathrm{d}s = \varepsilon\,\mathcal{A}(t)$$

where

$$\mathcal{A}(t) = \mathrm{D}_1\,H(u_\varepsilon(t), t/\varepsilon) - \mathrm{D}_1\,H(y, a/\varepsilon) \\ - \int_a^t \mathrm{D}_1\,H(u_\varepsilon(s), s/\varepsilon)\left[f(u_\varepsilon(s)) + h(u_\varepsilon(s), s/\varepsilon)\right]\mathrm{d}s\,.$$

Moreover,

$$u_\varepsilon(t) = y + \int_a^t f(u_\varepsilon(s))\,\mathrm{d}s + \int_a^t h(u_\varepsilon(s), s/\varepsilon)\,\mathrm{d}s$$

$$= y + \int_a^t f(u_\varepsilon(s))\,\mathrm{d}s + \varepsilon\,\mathcal{A}(t),$$

$$w(t) = y + \int_a^t f(w(s))\,\mathrm{d}s.$$

Hence

$$u_\varepsilon(t) - w(t) = \int_a^t \big[f(u_\varepsilon(s)) - f(w(s))\big]\,\mathrm{d}s + \varepsilon\,\mathcal{A}(t).$$

(3.13) and (3.16) imply that

$$\|f(u_\varepsilon(s)) - f(w(s))\| \le \mu\,\|u_\varepsilon(s) - w(s)\|,$$

$$\|\mathcal{A}(t)\| \le 2\nu + (b-a)\,\nu\,(\mu + \nu)$$

and

$$\|u_\varepsilon(t) - w(t)\| \le \int_a^t \mu\,\|u_\varepsilon(s) - w(s)\|\,\mathrm{d}s$$

$$+ \varepsilon\,[(2 + (b-a)\,\mu)\,\nu + (b-a)\,\nu^2].$$

Therefore (3.18) is correct by Lemma A.1. Moreover, (3.6) holds with

$$\varkappa_1 = (2 + \mu)\,(b-a)\,\mathrm{e}^{\mu(b-a)}, \quad \varkappa_2 = (b-a)\,\mathrm{e}^{\mu\,(b-a)}.$$

$\square$

3.4. Remark. Let $0 < r < R$, $y \in B(r)$,

$$\varepsilon\,\big[(2 + \mu)\,(b-a)\,\nu + (b-a)\,\nu^2\big]\,\mathrm{e}^{\mu\,(b-a)} \le R - r. \qquad (3.19)$$

Let there exist a solution $w : [a, b] \to B(r)$ of (3.2), $w(a) = y$. By classical results on ordinary differential equations u_ε exists on an interval $[a, c]$, $a < c \le b$ and can be continued to $[a, b]$ (cf. (3.18)) and (3.19)).

Chapter 4

Elementary methods: internal resonance

The solution w of (3.2) can be viewed as an approximation of the solutions u_ε of (3.1). A more precise approximation $\widetilde{w}_\varepsilon$ is constructed in this chapter but fairly stronger conditions are imposed on the right-hand side of (3.1).

4.1. Notation. Let $n, k \in \mathbb{N}$; $\varepsilon, R, \mu, \nu \in \mathbb{R}^+$.

Let $f : B(R) \to \mathbb{R}^n$, $h_i : B(R) \to \mathbb{R}^n$, $\lambda_i : \mathbb{R} \to \mathbb{R}$ for $i = 1, 2, \ldots, k$. Assume that f and h_i are differentiable,

$$
\left. \begin{aligned}
&\|f(x)\| \le \mu, \quad \|\mathrm{D}f(x)\| \le \mu, \\
&\|\mathrm{D}f(x) - \mathrm{D}f(\bar{x})\| \le \mu \, \|x - \bar{x}\| \quad \text{for } x, \bar{x} \in B(R),
\end{aligned} \right\} \tag{4.1}
$$

$$
\left. \begin{aligned}
&\|h_i(x)\| \le \nu \, k^{-1}, \quad \|\mathrm{D}h_i(x)\| \le \nu \, k^{-1}, \\
&\|\mathrm{D}h_i(x) - \mathrm{D}h_i(\bar{x})\| \le \nu \, k^{-1} \, \|x - \bar{x}\| \\
&\qquad\qquad \text{for } x, \bar{x} \in B(R), \quad i = 1, 2, \ldots, k,
\end{aligned} \right\} \tag{4.2}
$$

$$
\left. \begin{aligned}
&\lambda_i \quad \text{is continuous}, \\
&|\lambda_i(t)| \le 1, \ \lambda_i(t+1) = \lambda_i(t), \ \int_t^{t+1} \lambda_i(s) \, \mathrm{d}s = 0, \\
&\qquad\qquad \text{for } i = 1, 2, \ldots, k.
\end{aligned} \right\} \tag{4.3}
$$

By Lemma A.2 there exists $\Lambda_i : \mathbb{R} \to \mathbb{R}$ such that

$$
\left. \begin{aligned}
&\frac{\mathrm{d}\Lambda_i}{\mathrm{d}t}(t) = \lambda_i(t), \quad \Lambda_i(t+1) = \Lambda_i(t), \\
&\int_t^{t+1} \Lambda_i(t) \, \mathrm{d}\tau = 0, \quad |\Lambda_i(t)| \le \tfrac{1}{2}
\end{aligned} \right\} \tag{4.4}
$$

for $t \in \mathbb{R}$, $i = 1, 2, \dots, k$. Put

$$\xi_{i,j} = \int_0^1 \Lambda_i(\tau)\,\lambda_j(\tau)\,\mathrm{d}\tau, \quad i, j = 1, 2, \dots, k. \tag{4.5}$$

Observe that

$$|\xi_{i,j}| \le \tfrac{1}{2} \tag{4.6}$$

and that

$$\Lambda_i(t)\,\lambda_j(t) + \Lambda_j(t)\,\lambda_i(t) = \frac{\mathrm{d}}{\mathrm{d}t}\left(\Lambda_i(t)\,\Lambda_j(t)\right),$$

$$\int_0^1 \left(\Lambda_i(\tau)\,\lambda_j(\tau) + \Lambda_j(\tau)\,\lambda_i(\tau)\right)\mathrm{d}\tau$$

$$= \Lambda_i(1)\,\Lambda_j(1) - \Lambda_i(0)\,\Lambda_j(0) = 0.$$

Hence

$$\xi_{i,j} + \xi_{j,i} = 0, \quad i, j = 1, 2, \dots, k. \tag{4.7}$$

If $p, q : B(R) \to \mathbb{R}^n$ and if $\mathrm{D}p(x)$, $\mathrm{D}q(x)$ exist for $x \in B(R)$ then $\mathrm{D}p \circ q(x)$ is the composition of the maps $\mathrm{D}p$ and q,

$$\mathrm{D}p \circ q(x) = \left(\mathrm{D}p(x)\right)q(x) \tag{4.8}$$

and $[p, q](x)$ denotes the Lie bracket of the vector fields p, q, i.e.

$$\left.\begin{aligned}
[p, q](x) &= \mathrm{D}p \circ q(x) - \mathrm{D}q \circ p(x) \\
&= \mathrm{D}p(x)\,q(x) - \mathrm{D}q(x)\,p(x), \quad x \in B(R).
\end{aligned}\right\} \tag{4.9}$$

Let $y \in B(R)$, $[a, b] \subset \mathbb{R}$, $\varepsilon, \nu \in \mathbb{R}^+$. Assume that

$$0 < \varepsilon \le 1, \tag{4.10}$$

that $u_\varepsilon : [a, b] \to B(R)$ is a solution of

$$\dot{x} = f(x) + \sum_{i=1}^k h_i(x)\,\lambda_i(t/\varepsilon), \quad u_\varepsilon(a) = y \tag{4.11}$$

and $w_\varepsilon : [a, b] \to B(R - \tfrac{1}{4}\varepsilon\nu)$ is a solution of

$$\left.\begin{aligned}
\dot{x} &= f(x) - \varepsilon \sum_{i>j}[h_i, h_j](x)\,\xi_{i,j}, \\
w_\varepsilon(a) &= y - \varepsilon \sum_i h_i(y)\,\Lambda_i(a/\varepsilon).
\end{aligned}\right\} \tag{4.12}$$

Put

$$\widetilde{w}_\varepsilon(t) = w_\varepsilon(t) + \varepsilon \sum_i h_i(w_\varepsilon(t)) \Lambda_i(t/\varepsilon), \quad t \in [a,b]. \qquad (4.13)$$

Assume that

$$\widetilde{w}_\varepsilon(t) \in B(R) \quad \text{for } t \in [a,b]. \qquad (4.14)$$

Observe (cf. (4.2), (4.4)) that

$$\|\widetilde{w}_\varepsilon(t) - w_\varepsilon(t)\| \le \varepsilon \tfrac{1}{4}\nu, \quad t \in [a,b] \qquad (4.15)$$

which implies that $\widetilde{w}_\varepsilon(t) \in B(R)$ for $t \in [a,b]$. Finally, assume that

$$\varepsilon\,\nu^2 \le 1. \qquad (4.16)$$

The goal of this Chapter is to obtain an estimate for

$$\|u_\varepsilon(t) - \widetilde{w}_\varepsilon(t)\|, \quad t \in [a,b].$$

The crucial step is the inequality

$$\|u_\varepsilon(t) - \widetilde{w}_\varepsilon(t)\| \le \mathbf{C} + \mathbf{B} \int_a^t \|u_\varepsilon(s) - \widetilde{w}_\varepsilon(s)\|\, \mathrm{d}s, \quad t \in [a,b], \qquad (4.17)$$

where $\mathbf{C}$, $\mathbf{B}$ are positive functions of μ, ν, ε, t. (4.17) implies that

$$\|u_\varepsilon(t) - \widetilde{w}_\varepsilon(t)\| \le \exp\left(\int_a^t \mathbf{B}(s)\, \mathrm{d}s\right) \max\{\mathbf{C}(\tau); \tau \in [a,t]\}$$

for $t \in [a,b]$ (cf. Lemma A.1). Assumption (4.16) is introduced in order to keep $\displaystyle\int_a^t \mathbf{B}(s)\, \mathrm{d}s$ bounded. It will also make many formulas simpler.

4.2. Lemma. *Let $[\sigma, s] \subset [a,b]$. Then*

$$\|w_\varepsilon(s) - w_\varepsilon(\sigma)\| \le (\mu + \varepsilon\,\nu^2)\,(s - \sigma). \qquad (4.18)$$

Proof. (4.18) is a consequence of (4.12), (4.1), (4.2), (4.6), (4.7), (4.16) since

$$\sum_{i>j}[h_i, h_j](x)\,\xi_{i,j} = \sum_{i,j}\mathrm{D}\,h_i \circ h_j(x)\,\xi_{i,j}$$

and

$$\varepsilon \sum_{i,j} \|\mathrm{D}\,h_i \circ h_j(x)\|\,|\xi_{i,j}| \le \varepsilon\,\nu^2 \le 1.$$

$\square$

4.3. Lemma. *Let* $\rho \in \mathbb{R}^+$, $[\sigma, \sigma + \varepsilon] \subset [a, b]$ *and let* $p : B(R) \to \mathbb{R}^n$ *fulfil*

$$\|p(x)\| \le \rho, \|p(x) - p(\bar{x})\| \le \rho \|x - \bar{x}\| \quad for\ x,\ \bar{x} \in B(R). \tag{4.19}$$

Let $\Xi : \mathbb{R} \to \mathbb{R}$ *be continuous and fulfil*

$$|\Xi(t)| \le 1,\ \Xi(t + 1) = \Xi(t),\ \int_t^{t+1} \Xi(s)\,\mathrm{d}s = 0 \quad for\ t \in \mathbb{R}. \tag{4.20}$$

Then

$$\left\| \int_\sigma^{\sigma+\varepsilon} p\left(w_\varepsilon(s)\right) \Xi(s/\varepsilon)\,\mathrm{d}s \right\| \le \varepsilon^2\, \rho\,(\mu + 1). \tag{4.21}$$

Proof.

$$\int_\sigma^{\sigma+\varepsilon} p\left(w_\varepsilon(s)\right) \Xi(s/\varepsilon)\,\mathrm{d}s = \mathcal{J}(\sigma) + \mathcal{K}(\sigma),$$

where

$$\mathcal{J}(\sigma) - \int_\sigma^{\sigma+\varepsilon} \left[p\left(w_\sigma(s)\right) - p\left(w_\varepsilon(\sigma)\right)\right] \Xi(s/\varepsilon)\,\mathrm{d}s,$$

$$\mathcal{K}(\sigma) = \int_\sigma^{\sigma+\varepsilon} p\left(w_\varepsilon(\sigma)\right) \Xi(s/\varepsilon)\,\mathrm{d}s.$$

By (4.16), (4.18)

$$\|\mathcal{J}(\sigma)\| \le \int_\sigma^{\sigma+\varepsilon} \rho\,(\mu + 1)\,(s - \sigma)\,\mathrm{d}s \le \rho\,(\mu + 1)\,\varepsilon^2$$

and by (4.21) $\mathcal{K}(\sigma) = 0$, i.e. (4.20) holds. $\qquad\square$

4.4. Lemma. *Let* p *fulfil* (4.19) *and let* Ξ *fulfil* (4.20). *Then*

$$\left. \begin{aligned} \left\| \int_a^t p\left(w_\varepsilon(s)\right) \Xi(s/\varepsilon)\,\mathrm{d}s \right\| &\le \varepsilon\,\rho\left[(\mu + 1)\,(b - a) + 1\right] \\ &\qquad for\ a < t \le b. \end{aligned} \right\} \tag{4.22}$$

Proof.   Let $a < t \le b$ and let $m \in \mathbb{N}_0$ and $\sigma_\ell \in \mathbb{R}$ be defined by

$$m\varepsilon < t - a \le (m + 1)\varepsilon,\ \sigma_\ell = a + \ell\varepsilon \quad for\ \ell = 0, 1, \ldots, m. \tag{4.23}$$

Then

$$\int_a^t p\left(w_\varepsilon(s)\right) \Xi(s/\varepsilon)\mathrm{d}s$$

$$= \sum_{\ell=0}^{m-1} \int_{\sigma_\ell}^{\sigma_\ell+\varepsilon} p\left(w_\varepsilon(s)\right) \Xi(s/\varepsilon)\mathrm{d}s + \int_{\sigma_m}^t p\left(w_\varepsilon(s)\right) \Xi(s/\varepsilon)\mathrm{d}s.$$

By (4.21)

$$\left\| \int_{\sigma_\ell}^{\sigma_\ell+\varepsilon} p\left(w_\varepsilon(s)\right) \Xi(s/\varepsilon)\,\mathrm{d}s \right\| \leq \varepsilon^2 \rho\left(\mu+1\right).$$

Moreover (cf. (4.19), (4.20), (4.23)),

$$\left\| \int_{\sigma_m}^t p\left(w_\varepsilon(s)\right) \Xi(s/\varepsilon)\,\mathrm{d}s \right\| \leq \varepsilon \rho \quad \text{and} \quad m \leq \frac{b-a}{\varepsilon}.$$

Therefore (4.22) is correct. $\qquad\square$

4.5. Lemma. *Let* $p\colon B(R) \to \mathbb{R}^n$ *be differentiable and fulfil*

$$\left.\begin{array}{c} \|p\left(x\right)\| \leq \rho,\ \|\mathrm{D}p\left(x\right)\| \leq \rho,\ \|\mathrm{D}p\left(x\right) - \mathrm{D}p\left(\bar{x}\right)\| \leq \rho \\ \textit{for } x,\,\bar{x} \in B(R). \end{array}\right\} \tag{4.24}$$

Then

$$\left.\begin{array}{c} \left\| \int_a^t \left[p\left(\widetilde{w}_\varepsilon(s)\right) - p\left(w_\varepsilon(s)\right)\right] \right\| \\[2mm] \leq \varepsilon^2 \rho\left[(b-a+1)\,\nu + (b-a)\,\mu\,\nu + (b-a)\,\nu^2\right]. \end{array}\right\} \tag{4.25}$$

Proof. Let $a < t \leq b$. Then

$$\int_a^t \left[p\left(\widetilde{w}_\varepsilon(s) - p\left(w_\varepsilon(s)\right)\right)\right] \mathrm{d}s = \mathcal{L} + \mathcal{N}$$

where

$$\mathcal{L} = \int_a^t \left[p\left(\widetilde{w}_\varepsilon(s)\right) - p\left(w_\varepsilon(s)\right) - \mathrm{D}p\left(w_\varepsilon(s)\right)(\widetilde{w}_\varepsilon(s) - w_\varepsilon(s))\right]\mathrm{d}s,$$

$$\mathcal{N} = \int_a^t \mathrm{D}p\left(w_\varepsilon(s)\right)(\widetilde{w}_\varepsilon(s) - w_\varepsilon(s))\,\mathrm{d}s.$$

By (4.14), (4.15)

$$p\left(\widetilde{w}_\varepsilon(s)\right) - p\left(w_\varepsilon(s)\right)$$
$$= \int_0^1 \mathrm{D}p\left(w_\varepsilon(s) + \lambda\left(\widetilde{w}_\varepsilon(s) - w_\varepsilon(s)\right)\right)\left(\widetilde{w}_\varepsilon(s) - w_\varepsilon(s)\right)\mathrm{d}\lambda$$

and

$$p\left(\widetilde{w}_\varepsilon(s)\right) - p\left(w_\varepsilon(s)\right) - \mathrm{D}p\left(w_\varepsilon(s)\right)\right)\left(\widetilde{w}_\varepsilon(s) - w_\varepsilon(s)\right)$$
$$= \int_0^1 \left[\mathrm{D}p\left(w_\varepsilon(s) + \lambda[\widetilde{w}_\varepsilon(s) - w_\varepsilon(s)]\right) - \mathrm{D}p\left(w_\varepsilon(s)\right)\right]\left(\widetilde{w}_\varepsilon(s) - w_\varepsilon(s)\right)\mathrm{d}\lambda.$$

Hence

$$\left\|p\left(\widetilde{w}_\varepsilon(s)\right) - p\left(w_\varepsilon(s)\right) - \mathrm{D}p\left(w_\varepsilon(s)\right)\left(\widetilde{w}_\varepsilon(s) - w_\varepsilon(s)\right)\right\| \le \rho\left(\varepsilon\nu\right)^2$$

since (cf. (4.15)) $\left\|\widetilde{w}_\varepsilon(s) - w_\varepsilon(s)\right\| \le \varepsilon\nu$ and

$$\|\mathcal{L}\| \le \rho\left(b - a\right)(\varepsilon\nu)^2. \tag{4.26}$$

By (4.13)

$$\mathcal{N} = \sum_{i=1}^k \varepsilon \int_a^t \mathrm{D}p\left(w_\varepsilon(s)\right)h_i(w_\varepsilon(s))\Lambda_i(s/\varepsilon)\,\mathrm{d}s. \tag{4.27}$$

The integrals in (4.27) can be estimated similarly as in Lemma 4.4 since (cf. (4.2), (4.4))

$$\|\mathrm{D}p\left(x\right)h_i(x)\| \le \rho\nu k^{-1},$$

$$\|\mathrm{D}p\left(x\right)h_i(x) - \mathrm{D}p\left(\bar{x}\right)h_i(\bar{x})\| \le 2\rho\nu k^{-1}\|x - \bar{x}\| \quad \text{for } x, \bar{x} \in B(R),$$

and

$$|\Lambda_i(t)| \le \tfrac{1}{2} \text{ for } t \in \mathbb{R}.$$

Hence (cf. (4.4))

$$\left.\begin{aligned}\|\mathcal{N}\| &\le k\varepsilon\cdot\varepsilon\,2\,\rho\nu k^{-1}\left[(\mu + 1)\left(b - a\right) + 1\right]\tfrac{1}{2}\\ &\le \varepsilon^2\rho\left[(\mu + 1)\left(b - a\right) + 1\right]\nu.\end{aligned}\right\} \tag{4.28}$$

(4.25) is a consequence of (4.26) and (4.28). $\qquad\square$

4.6. Theorem. *There are functions* $\beta_j : (\mathbb{R}^+)^2 \to \mathbb{R}^+$, $j = 1, 2, 3$ *such that*

$$\left.\begin{aligned}\|u_\varepsilon(t) &- \widetilde{w}_\varepsilon(t)\|\\ &\le \varepsilon^2\left[\beta_1(b - a, \mu)\nu + \beta_2(b - a, \mu)\nu^2 + \beta_3(b - a, \mu)\nu^3\right]\end{aligned}\right\} \tag{4.29}$$

for $t \in [a, b]$.

Proof. By (4.11)

$$u_\varepsilon(t) = y + \int_a^t f(u_\varepsilon(s)) \, \mathrm{d}s + \sum_{i=1}^k \int_a^t h_i(u_\varepsilon(s)) \, \lambda_i(s/\varepsilon) \, \mathrm{d}s \,,$$

$$\int_a^t h_i(u_\varepsilon(s)) \, \lambda_i(s/\varepsilon) \, \mathrm{d}s = \varepsilon \, h_i(u_\varepsilon(t)) \Lambda_i(t/\varepsilon) - \varepsilon \, h_i(y) \Lambda_i(a/\varepsilon)$$

$$- \varepsilon \int_a^t \mathrm{D}\, h_i(u_\varepsilon(s)) \dot{u}_\varepsilon(s) \Lambda_i(s/\varepsilon) \mathrm{d}s \,,$$

$$\int_a^t \mathrm{D}h_i(u_\varepsilon(s)) \, \dot{u}_\varepsilon(s) \, \Lambda_i(s/\varepsilon) \, \mathrm{d}s = \int_a^t \mathrm{D}h_i(u_\varepsilon(s)) \, f(u_\varepsilon(s)) \, \Lambda_i(s/\varepsilon) \, \mathrm{d}s$$

$$+ \sum_j \int_a^t \mathrm{D}\, h_i(u_\varepsilon(s)) \, h_j(u_\varepsilon(s)) \, \Lambda_i(s/\varepsilon) \, \lambda_j(s/\varepsilon) \, \mathrm{d}s \,.$$

Hence

$$\left. \begin{aligned} u_\varepsilon(t) &= \varepsilon \sum_i h_i(u_\varepsilon(t)) \Lambda_i(t/\varepsilon) + y - \varepsilon \sum_i h_i(y) \, \Lambda_i(a/\varepsilon) \\ &+ \int_a^t f(u_\varepsilon(s)) \mathrm{d}s - \varepsilon \sum_i \int_a^t \mathrm{D}\, h_i \circ f(u_\varepsilon(s)) \, \Lambda_i(s/\varepsilon) \, \mathrm{d}s \\ &- \varepsilon \sum_{i,j} \int_a^t \mathrm{D}\, h_i \circ h_j(u_\varepsilon(s)) \, \Lambda_i(s/\varepsilon) \, \lambda_j(s/\varepsilon) \, \mathrm{d}s \,. \end{aligned} \right\} \quad (4.30)$$

By (4.12)

$$\left. \begin{aligned} w_\varepsilon(t) &= y - \varepsilon \sum_i h_i(y) \, \Lambda_i(a/\varepsilon) + \int_a^t f(w_\varepsilon(s)) \, \mathrm{d}s \\ &- \varepsilon \sum_{i,j} \int_a^t \mathrm{D}\, h_i \circ h_j(w_\varepsilon(s)) \, \xi_{i,j} \, \mathrm{d}s \end{aligned} \right\} \quad (4.31)$$

since (cf. (4.9), (4.7))

$$\sum_{i>j} [h_i, h_j](x) \, \xi_{i,j} = \sum_{i,j} \mathrm{D}\, h_i \circ h_j(x) \, \xi_{i,j} \,, \quad x \in B(R).$$

Subtracting (4.31) from (4.30) we obtain

$$u_\varepsilon(t) - w_\varepsilon(t)$$
$$= \varepsilon \sum_i h_i(u_\varepsilon(t)) \, \Lambda_i(t/\varepsilon) + \int_a^t \left[f(u_\varepsilon(s)) - f(w_\varepsilon(s)) \right] \mathrm{d}s$$

$$-\varepsilon \sum_i \int_a^t \mathrm{D}\, h_i \circ f(u_\varepsilon(s))\, \Lambda_i(s/\varepsilon)\, \mathrm{d}s$$

$$-\varepsilon \sum_{i,j} \int_a^t \mathrm{D}\, h_i \circ h_j(u_\varepsilon(s))\, \Lambda_i(s/\varepsilon)\, \lambda_j(s/\varepsilon)\, \mathrm{d}s$$

$$+\varepsilon \sum_{i,j} \int_a^t \mathrm{D}\, h_i \circ h_j(w_\varepsilon(s))\, \xi_{i,j}\, \mathrm{d}s,$$

which can be transformed (cf. also (4.13)) to

$$
\left.
\begin{aligned}
& u_\varepsilon(t) - \widetilde{w}_\varepsilon(t) = \int_a^t \big[f(u_\varepsilon(s)) - f(\widetilde{w}_\varepsilon(s)) \big]\, \mathrm{d}s \\[2mm]
& + \int_a^t \big[f(\widetilde{w}_\varepsilon(s)) - f(w_\varepsilon(s)) \big]\, \mathrm{d}s \\[2mm]
& -\varepsilon \sum_i \int_a^t \big[\mathrm{D}\, h_i \circ f(u_\varepsilon(s)) - \mathrm{D}\, h_i \circ f(\widetilde{w}_\varepsilon(s)) \big] \Lambda_i(s/\varepsilon)\, \mathrm{d}s \\[2mm]
& -\varepsilon \sum_i \int_a^t \big[\mathrm{D}\, h_i \circ f(\widetilde{w}_\varepsilon(s)) - \mathrm{D}\, h_i \circ f(w_\varepsilon(s)) \big] \Lambda_i(s/\varepsilon)\, \mathrm{d}s \\[2mm]
& -\varepsilon \int_a^t \mathrm{D}\, h_i \circ f(w_\varepsilon(s))\, \Lambda_i(s/\varepsilon)\, \mathrm{d}s \\[2mm]
& -\varepsilon \sum_{i,j} \int_a^t \big[\mathrm{D}\, h_i \circ h_j(u_\varepsilon(s)) - \mathrm{D}\, h_i \circ h_j(\widetilde{w}_\varepsilon(s)) \big] \Lambda_i(s/\varepsilon)\, \lambda_j(s/\varepsilon)\, \mathrm{d}s \\[2mm]
& -\varepsilon \sum_{i,j} \int_a^t \big[\mathrm{D}\, h_i \circ h_j(\widetilde{w}_\varepsilon(s)) - \mathrm{D}\, h_i \circ h_j(w_\varepsilon(s)) \big] \Lambda_i(s/\varepsilon)\, \lambda_j(s/\varepsilon)\, \mathrm{d}s \\[2mm]
& -\varepsilon \sum_{i,j} \int_a^t \mathrm{D}\, h_i \circ h_j(w_\varepsilon(s)) \big[\Lambda_i(s/\varepsilon)\, \lambda_j(s/\varepsilon) - \xi_{i,j} \big]\, \mathrm{d}s.
\end{aligned}
\right\} \quad (4.32)
$$

Now the terms on the right-hand side of (4.32) will be estimated. By (4.1)

$$\left\| \int_a^t \big[f(u_\varepsilon(s)) - f(\widetilde{w}_\varepsilon)) \big]\, \mathrm{d}s \right\| \le \mu \int_0^t \| u_\varepsilon(s) - \widetilde{w}_\varepsilon(s) \|\, \mathrm{d}s. \qquad (4.33)$$

By Lemma 4.5 with $p = f$, $\rho = \mu$

$$
\left.
\begin{aligned}
& \left\| \int_a^t \big[f(\widetilde{w}_\varepsilon(s)) - f(w_\varepsilon(s))) \big]\, \mathrm{d}s \right\| \\[2mm]
& \qquad \le \varepsilon^2 \big[((b-a+1)\, \mu + (b-a)\, \mu^2)\, \nu + (b-a)\, \mu \nu^2 \big].
\end{aligned}
\right\} \quad (4.34)
$$

Now, (4.1), (4.2), (4.4) and (4.10) imply that

$$\left\| \varepsilon \sum_i \int_a^t \left[\mathrm{D}\, h_i \!\circ\! f(u_\varepsilon(s)) - \mathrm{D}\, h_i \!\circ\! f(\widetilde{w}_\varepsilon(s)) \right] \Lambda_i(s/\varepsilon)\, \mathrm{d}s \right\|$$

$$\le \varepsilon\, 2\, \mu\, \nu \int_0^t \| u_\varepsilon(s) - \widetilde{w}_\varepsilon(s) \| \tfrac{1}{2}\, \mathrm{d}s \tag{4.35}$$

$$\le \mu \int_0^t \| u_\varepsilon(s) - \widetilde{w}_\varepsilon(s) \|\, \mathrm{d}s$$

since $\varepsilon\nu \le 1$ by (4.16) and (4.10). Similarly (cf. (4.15), (4.16))

$$\left\| \varepsilon \sum_i \int_a^t \left[\mathrm{D}h_i \!\circ\! f(\widetilde{w}_\varepsilon(s)) - \mathrm{D}h_i \!\circ\! f(w_\varepsilon(s)) \right] \Lambda_i(s/\varepsilon)\, \mathrm{d}s \right\|$$

$$\le \varepsilon\, 2\, \mu\, \nu \int_a^t \| \widetilde{w}_\varepsilon(s) - w_\varepsilon(s) \| \tfrac{1}{2}\, \mathrm{d}s \tag{4.36}$$

$$\le \varepsilon^2 (b-a)\, \mu\, \nu^2 .$$

Lemma 4.4 with $p = \sum_i \mathrm{D}\, h_i \circ f$, $\rho = 2\,\mu\,\nu$, $\Xi = \Lambda_i$ gives (cf. (4.4))

$$\left\| \varepsilon \sum_i \int_a^t \mathrm{D}\, h_i \circ f(w_\varepsilon(s)) \Lambda_i(s/\varepsilon)\, \mathrm{d}s \right\|$$

$$\le \varepsilon^2\, 2\, \mu\, \nu \left[(\mu+1)(b-a) + 1 \right] \tfrac{1}{2} \tag{4.37}$$

$$\le \varepsilon^2 \left[(b-a+1)\, \mu\, \nu + (b-a)\, \mu^2\, \nu \right] .$$

Further, (cf. (4.1), (4.2), (4.3), (4.4), (4.10), (4.16))

$$\left\| \varepsilon \sum_{i,j} \int_a^t \left[\mathrm{D}\, h_i \circ h_j(u_\varepsilon(s)) - \mathrm{D}\, h_i \circ h_j(\widetilde{w}_\varepsilon(s)) \right] \Lambda_i(s/\varepsilon)\, \lambda_j(s/\varepsilon)\, \mathrm{d}s \right\|$$

$$\le \varepsilon\, \nu^2\, 2 \int_a^t \| u_\varepsilon(s) - \widetilde{w}_\varepsilon(s) \|\, \mathrm{d}s \le \int_a^t \| u_\varepsilon(s) - \widetilde{w}_\varepsilon(s) \|\, \mathrm{d}s . \tag{4.38}$$

Similarly

$$\left\| \varepsilon \int_a^t \sum_{i,j} \left[\mathrm{D}\, h_i \circ h_j(\widetilde{w}_\varepsilon(t)) - \mathrm{D}\, h_i \circ h_j(w_\varepsilon(t)) \right] \Lambda_i(s/\varepsilon)\, \lambda_j(s/\varepsilon)\, \mathrm{d}s \right\|$$

$$\le \varepsilon\, (b-a)\, 2\, \nu^2\, \frac{\varepsilon\nu}{2} \le \varepsilon^2 (b-a)\, \nu^3 . \tag{4.39}$$

Finally, by Lemma 4.4 with

$$p = \mathrm{D}\, h_i \circ h_j, \quad \rho = 2\, \nu^2 k^{-2}, \quad \Xi = \Lambda_i\, \lambda_j - \xi_{i,j} ,$$

we get

$$\left\{\begin{aligned}
&\left\| \varepsilon \sum_{i,j} \int_a^t \mathbf{D}\, h_i \circ h_j(w_\varepsilon(s)) \left[\Lambda_i(s/\varepsilon)\, \lambda_j(s/\varepsilon) - \xi_{i,j} \right] \mathrm{d}s \right\| \\
&\qquad \leq \varepsilon^2\, 2\, \nu^2 \left[(\mu+1)\,(b-a) + 1 \right] \tfrac{1}{2} \\
&\qquad \leq \varepsilon^2 \left[(b-a+1) + (b-a)\,\mu \right] \nu^2
\end{aligned}\right\} \qquad (4.40)$$

since

$$|\lambda_j(t)| \leq 1, \quad |\Lambda_i(t)| \leq \tfrac{1}{2}, \quad |\xi_{i,j}| \leq \tfrac{1}{2},$$

$$|\Xi(t)| \leq 1, \ \ \Xi(t+1) = \Xi(t), \ \ \int_t^{t+1} \Xi(s)\,\mathrm{d}s = 0 \quad \text{for } t \in \mathbb{R}$$

and by Lemma A.2 there exists a primitive Θ of Ξ such that $\Theta(t+1) = \Theta(t)$ and $|\Theta(t)| \leq \tfrac{1}{4}$ for $t \in \mathbb{R}$.

Relations (4.32)–(4.40) imply that

$$\| u_\varepsilon(t) - \widetilde{w}_\varepsilon(t) \| \leq \mathbf{B} \int_a^t \| u_\varepsilon(s) - \widetilde{w}_\varepsilon(s) \| \,\mathrm{d}s + \mathbf{C}, \qquad (4.41)$$

where $\mathbf{B} = 2\,\mu + 1$ (cf. (4.33), (4.35), (4.38)). Moreover,

$$\mathbf{C} = \varepsilon^2 \left[\gamma_1(b-a,\mu)\,\nu + \gamma_2(b-a,\mu)\,\nu^2 + \gamma_3(b-a,\mu)\nu^3 \right],$$

where (cf. (4.34), (4.37))

$$\gamma_1(b-a,\mu) = 2\,(b-a+1)\,\mu + (b-a)\},\mu^2$$

and (cf. (4.34), (4.36), (4.40))

$$\gamma_2(b-a,\mu) = b-a+1 + 4\,(b-a)\,\mu$$

and (cf. (4.39))

$$\gamma_3 = b-a.$$

By Lemma A.2, (4.29) is true with

$$\beta_j(b-a,\mu) = \gamma_j(b-a,\mu)\exp\left((2\,\mu+1)\,(b-a) \right), \quad j = 1,2,3.$$

$\square$

4.7. Remark. The only bound on ν in Theorem 4.6 is $\varepsilon\,\nu^2 \le 1$ (cf. (4.16)). Let $0 \le \alpha \le \frac{1}{2}$, $\bar{\nu} \in \mathbb{R}^+$, $y \in B(R)$, let $\bar{h}_i : B(R) \to \mathbb{R}^n$ be differentiable and fulfil

$$\|\bar{h}(x)\| \le \bar{\nu}, \quad \|\mathrm{D}\bar{h}(x)\| \le \bar{\nu}, \quad \|\mathrm{D}\bar{h}(x) - \mathrm{D}\bar{h}(\bar{x})\| \le \bar{\nu}\,\|x - \bar{x}\|\,,$$

for x, $\bar{x} \in B(R)$. Theorem 4.6 can be applied to the equation

$$\dot{x} = f(x) + \sum_i \varepsilon^{-\alpha}\,\bar{h}(x)\,\lambda_i(t/\varepsilon)\,. \tag{4.42}$$

Let $u_\varepsilon : [a, b] \to B(R)$ be a solution of (4.42), $u_\varepsilon(a) = y$, let $w_\varepsilon : [a, b] \to B(R)$ be a solution of

$$\dot{x} = f(x) + \varepsilon^{1-2\alpha} \sum_{i>j} [\bar{h}_i, \bar{h}_j](x)\,,$$

$$w_\varepsilon(a) = y - \varepsilon^{1-\alpha} \sum_i h_i(y)\,\Lambda_i(a/\varepsilon)\,,$$

and

$$\widetilde{w}_\varepsilon(t) = w_\varepsilon(t) + \varepsilon^{1-\alpha} \sum_i h_i\,(w_\varepsilon(t))\,\Lambda_i(t/\varepsilon), \quad t \in [a, b]\,.$$

Theorem 4.6 implies that

$$\|u_\varepsilon(t) - \widetilde{w}_\varepsilon(t)\| \le \varepsilon^2\,[\beta_1(b - a, \mu)\,\varepsilon^{-\alpha}\,\bar{\nu} + \beta_2(b - a, \mu)\,\varepsilon^{-2\alpha}\,\bar{\nu}^2$$

$$+ \beta_3(b - a, \mu)\,\varepsilon^{-3\alpha}\,\bar{\nu}^3] \qquad \text{for } t \in [a, b]$$

and $u_\varepsilon(t) - \widetilde{w}_\varepsilon(t) \to 0$ uniformly on $[a, b]$ if $\alpha < \frac{2}{3}$.

4.8. Remark. Consider equation (4.11). Let u_ε, w_ε, $\widetilde{w}_\varepsilon$ have the same meaning as in Notation 4.1. Assume in addition to (4.1)–(4.3) that

$$a = 0, \quad \Lambda_i(a) = 0, \quad \xi_{i,j} = 0 \text{ for } i, j = 1, 2, \ldots, k\,. \tag{4.43}$$

Then w_ε is a solution of

$$\dot{x} = f(x), \quad w_\varepsilon(a) = y.$$

Hence it is independent of ε and we may write w instead of w_ε,

$$\widetilde{w}_\varepsilon(t) = w(t) + \varepsilon \sum_i h_i(w(t))\,\Lambda_i(t/\varepsilon) \tag{4.44}$$

and by Theorem 4.6 (cf. (4.1), (4.2), (4.16))

$$\left.\begin{aligned}
&\left\| u_\varepsilon(t) - w(t) - \varepsilon \sum_i h_i(w(t))\, \Lambda_i(t/\varepsilon) \right\| \\
&\qquad \leq \varepsilon^2 \left[\beta_1(b-a,\mu)\,\nu + \beta_2(b-a,\mu)\,\nu^2 + \beta_3(b-a,\mu)\,\nu^3 \right]
\end{aligned}\right\} \qquad (4.45)$$

for $\varepsilon \leq \nu^{-2}$.

Equation (2.2) which describes the motion of Kapitza's pendulum can be written in the form

$$\dot{x} = f(x) + h_1(x)\,\cos(4\pi t/\varepsilon) + h_2(x)\,\cos(2\pi t/\varepsilon) \qquad (4.46)$$

where

$$x = \begin{pmatrix} \Theta \\ \Phi \end{pmatrix}, \quad f(x) = \begin{pmatrix} \Phi \\ g\,L^{-1}\,\sin\Theta \end{pmatrix}, \quad \omega = 2\pi/\varepsilon,$$

$$h_1(x) = \begin{pmatrix} 0 \\ -\frac{1}{2}\,(L^{-1} A\, 2\pi/\varepsilon)^2\,\sin\Theta\,\cos\Theta \end{pmatrix},$$

$$h_2(x) = \begin{pmatrix} L^{-1} A\,(2\pi/\varepsilon)\,\sin\Theta \\ -L^{-1} A\,(2\pi/\varepsilon)\,\Phi\,\cos\Theta \end{pmatrix},$$

$$\lambda_1(t) = \cos 4\pi t, \quad \Lambda_1(t) = \frac{\sin 4\pi t}{4\pi},$$

$$\lambda_2(t) = \cos 2\pi t, \quad \Lambda_2(t) = \frac{\sin 2\pi t}{2\pi},$$

$$\xi_{i,j} = 0 \quad \text{for } i, j = 1, 2.$$

There exist R, μ, $\nu \in \mathbb{R}^+$ such that (4.1) and (4.2) hold. The estimate (4.45) is valid in the case of equation (4.46).

Chapter 5

Strong Riemann integration of functions of a pair of coupled variables

5.1. Notation. X is a Banach space, $\|x\|$ being the norm of $x \in X$, $\operatorname{Dom} U \subset \mathbb{R}^2$, $U : \operatorname{Dom} U \to X$.

Given $\sigma \in \mathbb{R}$ we denote by $U(\sigma, \cdot)$ the X-valued function which is defined for all $\tau \in \mathbb{R}$ such that $(\sigma, \tau) \in \operatorname{Dom} U$ with the value $U(\sigma, \tau)$ for such τ. Similarly for $U(\cdot, t)$ for a given $t \in \mathbb{R}$.

If $a, b \in \mathbb{R}$, $a \le b$, we put $[a, b] = \{t \in \mathbb{R} \,;\, a \le t \le b\}$. If $v : [a, b] \to X$, then

$$(\mathrm{R}) \int_a^b v(t) \, \mathrm{d}t$$

denotes the classical Riemann integral.

5.2. Definition. U is SR-*integrable* (*Strongly Riemann integrable*) on $[a, b]$ and u is an SR-*primitive* of U on $[a, b]$ if there exists $\xi_0 \in \mathbb{R}^+$ such that

$$(\tau, t) \in \operatorname{Dom} U \quad \text{for} \quad \tau, t \in [a, b], \quad \tau - \xi_0 \le t \le \tau + \xi_0 , \tag{5.1}$$

$$\left.\begin{array}{l} \text{for every } \varepsilon > 0 \text{ there exists } \xi > 0 \quad \text{such that} \\[2mm] \displaystyle\sum_{i=1}^{k} \| u(t_i) - u(t_{i-1}) - U(\tau_i, t_i) + U(\tau_i, t_{i-1}) \| \le \varepsilon \\[1mm] \text{for every set } A = (t_0, \tau_1, t_1, \tau_2, t_2, \ldots, \tau_k, t_k) \text{ fulfilling} \\[2mm] a = t_0 \le \tau_1 \le t_1 \le \tau_2 \le \ldots \le \tau_k \le t_k = b , \\[2mm] t_i - t_{i-1} \le \xi, \quad i = 1, 2, \ldots, k . \end{array}\right\} \tag{5.2}$$

Briefly, U is called SR-*integrable* and u is called *a primitive* of U, (τ, t) is called *a pair of coupled variables*.

5.3. Lemma. *Assume that U is SR-integrable on $[a,b]$, u is its primitive, $\varepsilon > 0$. Let $\xi > 0$ correspond to ε by Definition 5.2, $[S,T] \subset [a,b]$. Then*

$$\sum_{i=1}^{m} \|u(s_i) - u(s_{i-1}) - U(\sigma_i, s_i) + U(\sigma_i, s_{i-1})\| \leq \varepsilon \qquad (5.3)$$

for every sequence $A = (s_0, \sigma_1, s_1, \ldots, \sigma_m, s_m)$ fulfilling

$$\left. \begin{aligned} &S = s_0 \leq \sigma_1 \leq s_1 \leq \cdots \leq \sigma_m \leq s_m = T, \\ &s_i - s_{i-1} \leq \xi \quad \text{for } i = 1, 2, \ldots, m. \end{aligned} \right\} \qquad (5.4)$$

Proof.   Let the sequence A fulfil (5.4). There exists a sequence

$$B = (r_0, \rho_1, r_1, \ldots, \rho_n, r_n)$$

such that

$$a = r_0 \leq \rho_1 \leq r_1 \leq \cdots \leq \rho_n \leq r_n = S,$$
$$r_i - r_{i-1} \leq \xi \text{ for } i = 1, 2, \ldots, r$$

and a sequence $C = (\ell_0, \lambda_1, \ell_1, \ldots, \lambda_p, \ell_p)$ such that

$$T \leq \ell_0 \leq \lambda_1 \leq \ell - 1 \leq \ldots \lambda_p \leq \ell_p = b,$$
$$\ell_i - \ell_{i-1} \leq \xi \quad \text{for } i = 1, 2, \ldots, p.$$

Put

$$\begin{aligned} &t_i = r_i \quad \text{for } i = 0, 1, \ldots, n, \\ &\tau_i = \rho_i \quad \text{for } i = 1, 2, \ldots, n, \\ &t_i = s_{i-n}, \quad \tau_i = \sigma_{i-n} \qquad \text{for } i = n+1, n+2, \ldots, n+m, \\ &t_i = \ell_{i-n-m}, \quad \tau_i = \lambda_{i-n-m} \quad \text{for } i = n+m+1, n+m+2, \ldots, k, \end{aligned}$$

where $k = n + m + p$. Then

$$a = t_0 \leq \tau_1 \leq t_1 \leq \tau_2 \leq \cdots \leq \tau_k \leq t_k = b,$$
$$t_i - t_{i-1} \leq \xi, \quad i = 1, 2, \ldots, k.$$

Therefore by (5.2)

$$\sum_{j=1}^{m} \|u(s_j) - \mu(s_{j-1}) - U(\sigma_j, s_j) + U(\sigma_j, s_{j-1})\|$$

$$\leq \sum_{i=1}^{k} \|u(t_i) - u(t_{i-1}) - U(\tau_i, t_i) + U(\tau_i, t_{i-1})\| \leq \varepsilon$$

and (5.3) is correct. $\square$

5.4 . Corollary. *Let U be SR-integrable on $[a, b]$, u being its primitive, $[S, T] \subset [a, b]$. Then U is SR-integrable on $[S, T]$ and u is its primitive.*

5.5. Lemma. *Let U be SR-integrable and let u be a primitive of U on $[a, b]$.*

(i) *If $y \in X$ and $v(t) = u(t) + y$ for $t \in [a, b]$ then v is a primitive of U on $[a, b]$,*

(ii) *if v is another primitive of U on $[a, b]$ then*

$$u(T) - v(T) = u(a) - v(a) \quad \text{for } T \in [a, b].$$

Proof. (i) is a direct consequence of Definition 5.2. In order to prove (ii) assume that v is a primitive of U on $[a, b]$ and that $a < T \leq b$. By Corollary 5.4 U is SR-integrable on $[a, T]$, u and v being its primitives. Therefore there exists a sequence $A = (t_0, \tau_1, t_1, \ldots, \tau_k, t_k)$ such that

$$a = t_0 \leq \tau_1 \leq t_1 \leq \cdots \leq \tau_k \leq t_k = T,$$

$$\sum_{i=1}^{k} \|u(t_i) - u(t_{i-1}) - U(\tau_i, t_i) + U(\tau_i, t_{i-1})\| \leq \varepsilon,$$

$$\sum_{i=1}^{k} \|v(t_i) - v(t_{i-1}) - U(\tau_i, t_i) + U(\tau_i, t_{i+1})\| \leq \varepsilon.$$

Hence

$$\sum_{i=1}^{k} \|u(t_i) - u(t_{i-1}) - v(t_i) + v(t_{i-1})\| \leq 2\varepsilon. \tag{5.5}$$

Moreover,

$$u(T) - u(a) - v(T) + v(a) = \sum_{i=1}^{k} \big(u(t_i) - u(t_{i-1}) - v(t_i) + v(t_{i-1}) \big),$$

which together with (5.5) implies that

$$\|u(T) - v(T) - u(a) + v(a)\| \leq 2\varepsilon$$

and (ii) is true since $\varepsilon > 0$ is arbitrary. $\square$

5.6. Definition. Let U be SR-integrable and let u be its primitive on $[a, b]$. For $a \leq S \leq T \leq b$ put

$$(\text{SR}) \int_S^T D_t U(\tau, t) = u(T) - u(S). \tag{5.6}$$

The left-hand side of (5.6) is called the SR-*integral* of U over $[S, T]$.

Furthermore,

$$\text{(SR)} \int_T^S D_t U(\tau, t) = -\text{(SR)} \int_S^T D_t U(\tau, t), \quad \text{for } S < T \qquad (5.7)$$

and

$$\text{(SR)} \int_S^S D_t U(\tau, t) = 0. \qquad (5.8)$$

Observe that the left-hand side of (5.6) is uniquely defined since the right hand side of (5.6) does not depend on the choice of the primitive by Lemma 5.5.

5.7. Remark. Let $R, S, T \in [a, b]$. The preceding definition implies that

$$\text{(SR)} \int_R^T D_t U(\tau, t) = \text{(SR)} \int_R^S D_t U(\tau, t) + \text{(SR)} \int_S^T D_t U(\tau, t)$$

if U is SR-integrable on $[a, b]$ and u is its primitive.

5.8. Lemma. *Let* $u : [a, b] \to X$, $a < c < b$. *Assume that*

$$U \quad \text{is continuous at } (c, c), \qquad (5.9)$$
$$U \quad \text{is } SR\text{-integrable on } [a, c] \quad \text{and} \quad u \quad \text{is its primitive}, \qquad (5.10)$$
$$U \quad \text{is } SR\text{-integrable on } [c, b] \quad \text{and} \quad u \quad \text{is its primitive}. \qquad (5.11)$$

Then U *is SR-integrable on* $[a, b]$ *and* u *is its primitive.*

Proof. Let $\varepsilon > 0$. (5.7)–(5.9) imply that there exists $\xi \in \mathbb{R}$ such that $0 < \xi \leq \frac{1}{2}(b - a)$,

$$\left. \begin{aligned} &\|U(\tau, t) - U(c, c)\| \leq \varepsilon \\ &\quad \text{if } |\tau - c| \leq \xi, \ |t - c| \leq \xi, \ (\tau, t) \in \text{Dom} \, U, \end{aligned} \right\} \qquad (5.12)$$

$$\sum_{i=1}^k \|u(t_i) - u(t_{i-1}) - U(\tau_i, t_i) + U(\tau_i, t_{i-1})\| \leq \varepsilon \qquad (5.13)$$

for any sequence $(t_0, \tau_1, t_1, \ldots, t_k)$ fulfilling

$$\left. \begin{aligned} a = t_0 \leq \tau_1 \leq t_1 \leq \ldots \leq \tau_k \leq t_k = c, \ |t_i - t_{i-1}| \leq \xi \\ \text{for } i = 1, 2, \ldots, k \end{aligned} \right\} \qquad (5.14)$$

and

$$\sum_{i=1}^{\ell} \|u(s_i) - u(s_{i-1}) - U(\sigma_i, s_i) + U(\sigma_i, s_{i-1})\| \leq \varepsilon \tag{5.15}$$

for any sequence $(s_0, \sigma_1, s_1, \ldots, \sigma_\ell, s_\ell)$ fulfilling

$$\left. \begin{array}{r} c = s_0 \leq \sigma_1 \leq s_1 \leq \ldots \leq \sigma_\ell \leq s_\ell = b, \ |s_i - s_{i-1}| \leq \xi \\[2mm] \text{for } i = 1, 2, \ldots, \ell. \end{array} \right\} \tag{5.16}$$

Let the set $\{r_0, \rho_1, r_1, \ldots, r_m, \rho_m\}$ fulfil

$$\left. \begin{array}{r} a = r_0 \leq \rho_1 \leq r_1 \leq \ldots \leq \rho_m \leq r_m = b, \ |r_i - r_{i-1}| \leq \xi \\[2mm] \text{for } i = 1, 2, \ldots, m. \end{array} \right\} \tag{5.17}$$

Then $m > 2$ and either

$$\text{there is } p \text{ such that } 2 \leq p \leq m-2 \text{ and } r_p = c, \ r_{p+1} = c \tag{5.18}$$

or

$$\text{there is } p \geq 1 \text{ such that } r_{p-1} < c < r_p. \tag{5.19}$$

If (5.18) holds then (cf. (5.11), (5.12), (5.14), (5.16))

$$\left. \begin{array}{l} \displaystyle\sum_{i=1}^{m} \|u(r_i) - u(r_{i-1}) - U(\rho_i, r_i) + U(\rho_i, r_{i-1})\| \\[4mm] = \displaystyle\sum_{i=1}^{p} \|u(r_i) - u(r_{i-1}) - U(\rho_i, r_i) + U(\rho_i, r_{i-1})\| \\[4mm] + \displaystyle\sum_{j=p+1}^{m} \|u(r_j) - u(r_{j-1}) - U(\rho_j, r_j) + U(\rho_j, r_{j-1})\| \\[4mm] \leq 2\varepsilon. \end{array} \right\} \tag{5.20}$$

Let (5.19) take place. Then

$$\sum_{i=1}^{m} \|u(r_i) - u(r_{i-1}) - U(\rho_i, r_i) + U(\rho_i, r_{i-1})\| = K_1 + K_2 + K_3 \tag{5.21}$$

where

$$K_1 = \sum_{i=1}^{p-1} \|u(r_i) - u(r_{i-1}) - U(\rho_i, r_i) + U(\rho_i, r_{i-1})\|$$

$$+ \|u(c) - u(r_{p-1}) - U(c,c) + u(c, r_{p-1})\|$$

$$+ \|u(r_p) - u(c) - U(c, r_p) + U(c,c)\|$$

$$+ \sum_{i=p+1}^{m} \|u(r_i) - u(r_{i-1}) - U(\rho_i, r_i) + U(\rho_i, r_{i-1})\|,$$

$$K_2 = \|u(r_p) - u(r_{p-1}) - U(\rho_p, r_p) + u(\rho_p, r_{p-1})\|$$

$$- \|u(r_p) - u(r_{p-1}) - U(c, r_p) + U(c, r_{p-1})\|,$$

$$K_3 = \|u(r_p) - u(r_{p-1}) - U(c, r_p) + U(c, r_{p-1})\|$$

$$- \|u(c) - u(r_{p-1}) - U(c,c) + U(c, r_{p-1})\|$$

$$\|u(r_p) - u(c) - U(c, r_p) + U(c,c)\|.$$

By Definition 5.2

$$K_1 \leq \varepsilon \qquad (5.22)$$

since the sequence

$$\{t_0, \tau_1, t_1, \ldots, \tau_{p-1}, t_{p-1}, c, c, c, t_p, \tau_{p+1}, \ldots, \tau_m, t_m\}$$

fulfils (5.15). Further,

$$\|u(r_p) - u(r_{p-1}) - U(\rho_p, r_p) + U(\rho_p, r_{p-1})\|$$

$$\leq \|u(r_p) - u(r_{p-1}) - U(c, r_p) + U(c, r_{p-1})\|$$

$$+ \|U(\rho_p, r_p) - U(\rho_p, c_p)\| + \|U(\rho_p, r_{p-1}) - U(c, r_{p-1})\|$$

and by (5.7)

$$K_2 \leq 2\varepsilon. \qquad (5.23)$$

Finally,

$$\|u(r_p) - u(r_{p-1}) - U(c, r_p) + U(c, r_{p-1})\|$$

$$\leq \|u(r_p) - u(c) - U(c, r_p) + U(c, c)\|$$

$$+ \|u(c) - u(r_{p-1}) - U(c, c) + U(c, r_{p-1})\|.$$

Hence

$$K_3 \leq 0. \tag{5.24}$$

Putting (5.19)–(5.22) together we obtain that

$$\sum_{i=1}^{m} \|u(r_i) - u(r_{i-1}) - U(\rho_i, r_i) + U(\rho_i, r_{i-1})\| \leq 3\varepsilon. \tag{5.25}$$

(5.18) and (5.23) imply that U is SR-integrable on $[a, b]$ and u is its primitive. $\qquad \square$

5.9. Lemma. *Let U be SR-integrable on $[a, b]$, u being its primitive. Let $\sigma \in [a, b]$. Assume that*

$$U(\sigma, \cdot) : [\sigma - \xi_0, \sigma + \xi_0] \cap [a, b] \to X$$

is continuous at σ. Then u is continuous at σ.

Proof. Let $\varepsilon > 0$ and let $\xi > 0$ correspond to ε by Definition 5.2. Let $a \leq \bar{s} \leq \sigma \leq s \leq c$, $s - \bar{s} \leq \xi$. There exists $A = \{t_0, \tau_1, t_1, \ldots, \tau_k, t_k\}$ such that

$$a = t_0 \leq \tau_1 \leq t_1 \leq \tau_2 \leq \cdots \leq \tau_k \leq t_k = b,$$

$$t_i - t_{i-1} \leq \xi \quad \text{for} \quad i = 1, 2, \ldots, k$$

is fulfilled and $(\bar{s}, \sigma, s) = (t_{i-1}, \tau_i, t_i)$ for some $i = 1, 2, \ldots, k$. Then

$$\|u(s) - u(\bar{s}) - U(\sigma, s) + U(\sigma, \bar{s})\| \leq \varepsilon$$

by (5.2). $\qquad \square$

5.10. Remark. Let $w : [a, b] \to X$, $U(\tau, t) = w(\tau) t$ for $\tau, t \in [a, b]$. If U is SR-integrable and if u is its primitive then $(\text{R}) \int_a^b w(t) \, dt = u(b) - u(a)$, $(\text{R}) \int_a^b w(t) \, dt$ denoting the Riemann integral of w over $[a, b]$. This is a consequence of Definition 5.2.

On the other hand, if $(\mathrm{R})\displaystyle\int_a^b w(t)\,\mathrm{d}t$ exists, $y \in X$ and if

$$u(t) = (\mathrm{R}) \int_a^t w(s)\,\mathrm{d}s + y \quad \text{for } t \in [a,b]$$

then for every $\varepsilon > 0$ there exists $\xi > 0$ such that

$$\left\| \sum_{i=1}^k (u(t_i) - u(t_{i-1})) - U(\tau, t_i) + U(\tau, t_{i-1}) \right\| \leq \varepsilon$$

if $A = (t_0, \tau_1, t_1, \tau_2, t_2, \ldots, \tau_k, t_k)$ is a set fulfilling

$$a = t_0 \leq \tau_1 \leq t_1 \leq \tau_2 \leq \cdots \leq \tau_k \leq t_k = b$$

and

$$t_i - t_{i-1} \leq \xi, \quad i = 1, 2, \ldots, k,$$

but U need not be SR-integrable.

If $\dim X < \infty$ and if $(\mathrm{R})\displaystyle\int_a^b w(t)\,\mathrm{d}t$ exists then U is SR-integrable and u is its primitive. This is proved in Chapter 18 for the SKH-integration and it can be proved for the SR-integration in the same way (see also [Schwabik, Ye (2005)], Corollary 3.4.3 and Remark).

5.11. Remark. If $v : [a,b] \to X$ is continuous then the Riemann integral $(\mathrm{R})\displaystyle\int_a^b v(s)\,\mathrm{d}s$ exists. Furthermore, observe that

$$\left\| (\mathrm{R}) \int_{\bar{t}}^t v(s)\,\mathrm{d}s - v(\tau)(t - \bar{t}) \right\| \leq \varepsilon\,(t - \bar{t})$$

if $a \leq \bar{t} \leq \tau \leq t \leq b$ and if $\|v(s) - v(\tau)\| \leq \varepsilon$ for $s \in [\bar{t}, t]$.

5.12. Lemma. *Let* $v : [a,b] \to X$ *be continuous,* $U(\tau, t) = v(\tau)\,t$,

$$w(t) = (\mathrm{R}) \int_a^t v(s)\,\mathrm{d}s \quad \text{for } t \in [a,b].$$

Then

$$U \quad \text{is SR-integrable}, \quad w \text{ is its primitive}, \tag{5.26}$$

$$\frac{\mathrm{d}w}{\mathrm{d}t}(t) = v(t) \quad \text{for } t \subset [a,b]. \tag{5.27}$$

Proof.   For every $\varepsilon > 0$ there exists $\xi > 0$ such that

$$\|w(T) - w(S) - v(\tau)(T - S)\| \leq \varepsilon(T - S) \tag{5.28}$$

if $a \leq S \leq \tau \leq T \leq b$, $T - S \leq \xi$ since

$$w(T) - w(S) - v(\tau)(T - S) = (\mathrm{R}) \int_S^T [v(t) - v(\tau)]\, dt\,.$$

Both (5.26) and (5.27) are consequences of (5.28). $\square$

Chapter 6

Generalized ordinary differential equations: Strong Riemann-solutions (concepts)

6.1. Notation. $B(r) = \{x \in X \, ; \, \|x\| \leq r\}$ for $r \geq 0,$ $\operatorname{Dom} G \subset X \times \mathbb{R}^2,$
$G : \operatorname{Dom} G \to X,$ $[a, b] \subset \mathbb{R},$ $u : [a, b] \to X,$ $\xi_0 > 0,$

$$(u(\tau), \tau, t) \in \operatorname{Dom} G \text{ for } \tau, t \in [a, b], \quad \tau - \xi_0 \leq t \leq \tau + \xi_0 \,.$$

6.2. Definition. u is an SR-*solution* of the generalized differential equation (GODE)

$$\frac{\mathrm{d}}{\mathrm{d}t} x = \mathrm{D}_t G(x, \tau, t) \tag{6.1}$$

on $[a, b]$ if

$$u(T) - u(S) = (\mathrm{SR}) \int_S^T \mathrm{D}_t G(u(\tau), \tau, t) \quad \text{for } [S, T] \subset [a, b] \,. \tag{6.2}$$

Briefly, u is an SR-*solution* of (6.1).

6.3. Lemma. *Let* $u : [a, b] \to X,$ $P \in [a, b].$ *Then*

$$u \quad \text{is an } \mathrm{SR} - solution \text{ of } (6.1) \text{ on } [a, b] \tag{6.3}$$

if and only if

$$u(T) = u(P) + (\mathrm{SR}) \int_P^T \mathrm{D}_t G(u(\tau), \tau, t) \quad \text{for } T \in [a, b] \,. \tag{6.4}$$

Proof.   Let (6.3) hold. Then

$$u(T) = u(a) + (\text{SR}) \int_a^T D_t G(u(\tau), \tau, T) \quad \text{for} \ \ T \in [a, b],$$

$$u(P) = u(a) + (\text{SR}) \int_a^P D_t G(u(\tau), \tau, T).$$

Hence (cf. Remark 5.7)

$$u(T) = u(P) + (\text{SR}) \int_P^T D_t G(u(\tau), \tau, t) \quad \text{if} \ \ P \leq T \leq b,$$

$$u(T) = u(P) - (\text{SR}) \int_T^P D_t G(u(\tau), \tau, t) \quad \text{if} \ \ a \leq T \leq P,$$

and (6.4) holds by (5.7).

Let (6.4) hold. Then

$$u(S) = u(P) + (\text{SR}) \int_P^S D_t G(u(\tau), \tau, t) \quad \text{tor} \ \ S \in [a, b].$$

By (5.7) and Remark 5.7

$$u(T) - u(S) = (\text{SR}) \int_P^T D_t G(u(\tau), \tau, t) - (\text{SR}) \int_P^S D_t G(u(\tau), \tau, t)$$

$$= (\text{SR}) \int_P^T D_t G(u(\tau), \tau, t) + (\text{SR}) \int_S^P D_t G(u(\tau), \tau, t)$$

$$= (\text{SR}) \int_S^T D_t G(u(\tau), \tau, t)$$

and (6.3) is correct.   □

6.4. Lemma. *The two assertions below are equivalent:*

$\quad u \quad$ *is an SR-solution of* (6.1),   (6.5)

and

for every $\varepsilon > 0$ *there exists* $\xi > 0$ *such that*

$$\sum_{\mathcal{A}} \|u(t_i) - u(t_{i-1}) - G(u(\tau), \tau, t_i) + G(u(\tau), \tau, t_{i-1})\| \le \varepsilon$$

for every $A = (t_0, \tau_1, t_1, \tau_2, t_2, \ldots, \tau_k, t_k)$ *such that*

$$a = t_0 \le \tau_1 \le t_1 \le \tau_2 \le \ldots \le \tau_k \le t_k = b,$$

$$t_i - t_{i-1} \le \xi \quad for \quad i = 1, 2, \ldots, k.$$

(6.6)

Proof. Definition 5.2 implies that (6.2) and (6.6) are equivalent.   □

6.5 . Lemma. *Let u be an SR-solution of* (6.2) *on* $[a, b]$. *Assume that* $\sigma \in [a, b]$ *and that* $G(u(\sigma), \sigma, \cdot)$ *is continuous at* σ. *Then u is continuous at* σ.

Proof. This is a consequence of Definition 6.2 and Lemma 5.9.   □

6.6. Remark. Let u be an SR-solution of (6.1) on $[a, b]$. By Definition 6.2 u is an SR-solution of (6.1) on any $[c, d] \subset [a, b]$.

On the other hand, let $a < c < b$ and let u be an SR-solution of (6.2) on $[a, c]$ and on $[c, b]$. Assume that G is continuous at $(u(c), c, c)$. Then u is an SR-solution of (6.2) on $[a, b]$ by Lemma 5.8 since

(i) u is continuous at c by Lemma 6.5 ,

(ii)   if $U(\tau, t) = G(u(\tau), \tau, t)$ then U is continuous at (c, c).

6.7 . Lemma. *Let u be an SR-solution of* (6.1) *on* $[a, b]$, $[S, T] \subset [a, b]$. *Then u is an SR-solution of* (6.1) *on* $[S, T]$.

Proof. Lemma 6.7 is a consequence of Corollary 5.4.   □

6.8 . Definition. Let $\mathrm{Dom}\, g \subset B(r) \times \mathbb{R}^2$, $g : \mathrm{Dom}\, g \to X$. A function u is a *classical solution* of

$$\dot{x} = g(x, t, t) \tag{6.7}$$

on $[a, b]$ if

$$(u(\tau), \tau, \tau) \in \mathrm{Dom}\, g \quad for \quad \tau \in [a, b]$$

and if

$$\frac{du}{dt}(t) = g(u(t), t, t) \quad for \ t \in [a, b] . \tag{6.8}$$

6.9. Lemma. *Let* $[a, b] \subset \mathbb{R}$, $r > 0$, $g: B(r) \times [a, b]^2 \to X$ *and* $u: [a, b] \to B(r)$. *Let* g *and* u *be continuous. Put*

$$\operatorname{Dom} G = B(r) \times [a, b]^2,$$

and

$$G(x, \tau, t) = (\mathrm{R}) \int_a^t g(x, \tau, s) \, \mathrm{d}s \quad for \ (x, \tau, t) \in \operatorname{Dom} G. \tag{6.9}$$

Then

$$(\mathrm{SR}) \int_S^T \mathrm{D}_t G(u(\tau), \tau, t) = (\mathrm{R}) \int_S^T g(u(s), s, s) \, \mathrm{d}s. \tag{6.10}$$

Proof. Since g and u are continuous the integrals in (6.9), (6.10) exist. Let $\varepsilon > 0$. There exists $\xi > 0$ such that

$$\|g(u(\tau), \tau, s) - g(u(\tau), \tau, \bar{s})\| \leq \varepsilon, \tag{6.11}$$

$$\|g(u(s), s, s) - g(u(\bar{s}), \bar{s}, \bar{s})\| \leq \varepsilon \tag{6.12}$$

for $\tau, s, s \in [a, b]$, $|s - \bar{s}| \leq \zeta$.

Moreover, (6.11) implies (cf. Remark 5.11) that

$$\left\| (\mathrm{R}) \int_{\bar{t}}^t g(u(\tau), \tau, s) ds - g(u(\tau), \tau, \tau)(t - \bar{t}) \right\| \leq \varepsilon (t - \bar{t})$$

$$\text{if } a \leq \bar{t} \leq \tau \leq t \leq b \text{ and } t - \bar{t} \leq \xi.$$

Hence

$$\left\| G(u(\tau), \tau, t) - G(u(\tau), \tau, \bar{t}) - g(u(\tau), \tau, \tau)(t - \bar{t}) \right\|$$
$$= \left\| (\mathrm{R}) \int_{\bar{t}}^t g(u(\tau), \tau, s) \, \mathrm{d}s - g(u(\tau), \tau, \tau)(t - \bar{t}) \right\| \leq \varepsilon (t - \bar{t}) \left.\begin{array}{c} \\ \\ \\ \end{array}\right\} \tag{6.13}$$
$$\text{if } a \leq \bar{t} \leq \tau \leq t \leq b \text{ and } t - \bar{t} \leq \xi.$$

Further, by (6.12)

$$\left\| (\mathrm{R}) \int_{\bar{t}}^t g(u(s), s, s) ds - g(u(\tau), \tau, \tau)(t - \bar{t}) \right\| \leq \varepsilon (t - \bar{t}) \left.\begin{array}{c} \\ \\ \end{array}\right\} \tag{6.14}$$
$$\text{if } a \leq \bar{t} \leq \tau \leq t \leq b \text{ and } t - \bar{t} \leq \xi.$$

Relations (6.13), (6.14) imply that

$$\left\| G(u(\tau), \tau, t) - G(u(\tau), \tau, \bar{t}) - (\mathrm{R}) \int_{\bar{t}}^{t} g(u(s), s, s) \, \mathrm{d}s \right\| \leq 2\,\varepsilon\,(t - \bar{t})$$

$$\text{if } t - \bar{t} \leq \xi$$

and finally (cf. Lemma 6.4), $G(u(\tau), \tau, t)$ is SR-integrable on $[a, b]$, $(\mathrm{R}) \int_{a}^{t} g(u(s), s, s) \, \mathrm{d}s$ is its SR-primitive and (6.10) holds. □

6.10. Theorem. *Assume that* $[a, b] \subset \mathbb{R}$, $r > 0$, $g : B(r) \times [a, b]^2 \to X$ *and* $u : [a, b] \to B(r)$. *Let* g *and* u *be continuous and let* G *be defined by* (6.9). *Then*

$$u \quad \text{is a classical solution of (6.7)} \tag{6.15}$$

if and only if

$$u \text{ is an SR-solution of (6.1).}$$

Proof. If u is an SR-solution of (6.1) then

$$u(T) - u(S) = (\mathrm{R}) \int_{S}^{T} g(u(s), s, s) \, \mathrm{d}s \quad \text{for } [S, T] \subset [a, b] \tag{6.16}$$

by Lemma 6.9 and u is a classical solution of (6.7) since g and u are continuous (cf. Lemma 5.12).

If u is a classical solution of (6.7) then (6.16) holds again (cf. Remark 5.11) and u is an SR-solution of (6.1) by Lemma 6.9. □

Chapter 7

Functions ψ_1, ψ_2

7.1. Notation. Let ψ_1, $\psi_2 : \mathbb{R}_0^+ \to \mathbb{R}_0^+$. Assume that

$$\psi_1, \psi_2 \quad \text{are nondecreasing, continuous,} \tag{7.1}$$

$$\psi_1(\sigma) \le \psi_2(\sigma) \quad \text{for } \sigma \in \mathbb{R}_0^+ , \tag{7.2}$$

$$\sum_{i=1}^{\infty} 2^i \, \psi_1(2^{-i}) \, \psi_2(2^{-i}) < \infty , \tag{7.3}$$

$$\sum_{i=1}^{\infty} \psi_2(2^{-i}) < \infty . \tag{7.4}$$

Put

$$\Psi(\sigma) = \sum_{i=1}^{\infty} 2^i \psi_1(\sigma 2^{-i}) \, \psi_2(\sigma 2^{-i}) \quad \text{for } \sigma \in \mathbb{R}_0^+ , \tag{7.5}$$

$$\mathcal{A}(\sigma) = \Psi(\sigma) \exp \sum_{j=1}^{\infty} \psi_2(\sigma 2^{-j}) \quad \text{for } \sigma \in \mathbb{R}_0^+ , \tag{7.6}$$

$$\varphi(\sigma) = 3 \, \psi_1(\sigma) \, \psi_2(\sigma) + \mathcal{A}(\sigma) \, \psi_1(\sigma) \quad \text{for } \sigma \in \mathbb{R}_0^+ , \tag{7.7}$$

$$\Phi(\sigma) = \sum_{i=1}^{\infty} 2^{i-1} \, \varphi(\sigma 2^{-i}) \quad \text{for } \sigma \in \mathbb{R}_0^+ , \tag{7.8}$$

$$\mathcal{B}(\sigma) = \Phi(\sigma) \exp \left(\sum_{j=1}^{\infty} 2 \, \psi_2(\sigma 2^{-j}) + \sigma \right) \quad \text{for } \sigma \in \mathbb{R}_0^+ . \tag{7.9}$$

Observe that $\Psi, \mathcal{A}, \varphi$ are well defined (cf. (7.1)–(7.4)) but it remains to be proved that $\Phi, \mathcal{B}$ are well defined as well.

7.2. Remark. Let $\alpha > \frac{1}{2}$, $0 < \beta \leq \alpha$, $\alpha + \beta > 1$, $\psi_1(\sigma) = \sigma^\alpha$, $\psi_2(\sigma) = \sigma^\beta$ for $0 \leq \sigma \leq 1$, $\psi_1(\sigma) = \psi_2(\sigma) = \sigma$ for $\sigma \geq 1$. Then (7.1)–(7.4) hold and

$$\Psi(\sigma) = \sigma^{\alpha+\beta} \frac{2^{-\alpha-\beta+1}}{1 - 2^{-\alpha-\beta+1}},$$

$$\mathcal{A}(\sigma) = \Psi(\sigma) \exp\left(\sigma^\beta \frac{2^{-\beta}}{1 - 2^{-\beta}}\right) \quad \text{for } 0 \leq \sigma \leq 1.$$

7.3. Remark. (7.1)–(7.4) imply that Ψ, $\mathcal{A}$, φ are continuous and nondecreasing. Moreover,

$$\sum_{i=1}^{\infty} 2^i \, \Psi(\sigma\, 2^{-i}) \, \psi_1(\sigma 2^{-i})$$

$$= \sum_{i=1}^{\infty} 2^i \sum_{j=1}^{\infty} 2^j \, \psi_1(\sigma 2^{-i-j}) \, \psi_2(\sigma 2^{-i-j}) \, \psi_1(\sigma 2^{-i})$$

$$= \sum_{k=2}^{\infty} 2^k \, \psi_1(\sigma 2^{-k}) \, \psi_2(\sigma 2^{-k}) \sum_{i-1}^{k-1} \psi_1(\sigma 2^{-i}) < \infty$$

since $\displaystyle\sum_{i=1}^{\infty} \psi_1(\sigma\, 2^{-i}) < \infty$ by (7.2) and (7.4). Hence

$$\left.\begin{array}{l} \displaystyle\sum_{i=1}^{\infty} 2^i \, \mathcal{A}(\sigma 2^{-i}) \, \psi_1(\sigma 2^{-i}) \\[2mm] \displaystyle\leq \sum_{i=1}^{\infty} 2^i \Psi(\sigma 2^{-i}) \, \psi_1(\sigma 2^{-i}) \exp \sum_{j=1}^{\infty} \psi_1(\sigma\, 2^{-i-j}) < \infty \end{array}\right\} \tag{7.10}$$

and Φ and $\mathcal{B}$ are well defined, nondecreasing and continuous as well.

7.4. Lemma. Let $\lambda : \mathbb{R}_0^+ \to \mathbb{R}_0^+$ be nondecreasing, continuous,

$$\sum_{i=1}^{\infty} 2^i \, \lambda(2^{-i}) < \infty, \tag{7.11}$$

$$\Lambda(\sigma) = \sum_{i=1}^{\infty} 2^i \, \lambda(\sigma\, 2^{-i}) \quad \text{for } \sigma \in \mathbb{R}_0^+. \tag{7.12}$$

Then

$$\Lambda \quad \text{is nondecreasing, continuous,} \tag{7.13}$$

$$\lim_{\sigma \to 0+} \frac{1}{\sigma} \Lambda(\sigma) = 0. \tag{7.14}$$

Proof. The series in (7.11) is uniformly convergent on $[0, S]$ for $0 < S$. Hence $\mathcal{A}$ is continuous and nondecreasing. Let $k \in \mathbb{N}$. Then

$$2^k \Lambda(2^{-k}) = \sum_{i=k+1}^{\infty} 2^i \lambda(2^{-i}) \quad \text{and} \quad \lim_{k \to \infty} 2^k \Lambda(2^{-k}) = 0.$$

Moreover,

$$\frac{1}{\sigma} \Lambda(\sigma) \le 2^{k+1} \Lambda(2^{-k}) \quad \text{for } 2^{-k-1} \le \sigma \le 2^{-k}$$

and (7.14) is valid. $\qquad\qquad\qquad\qquad\qquad\qquad\qquad\qquad\qquad\qquad\square$

7.5. Remark. Lemma 7.4 applies if $\lambda(\sigma) = \psi_1(\sigma)\,\psi_2(\sigma)$. Therefore

$$\lim_{\sigma \to 0+} \frac{1}{\sigma} \Psi(\sigma) = 0, \tag{7.15}$$

$$\lim_{\sigma \to 0+} \frac{1}{\sigma} \mathcal{A}(\sigma) = 0. \tag{7.16}$$

Moreover (cf. (7.10)),

$$\lim_{\sigma \to 0+} \frac{1}{\sigma} \Phi(\sigma) = 0, \tag{7.17}$$

$$\lim_{\sigma \to 0+} \frac{1}{\sigma} \mathcal{B}(\sigma) = 0. \tag{7.18}$$

Observe that

$$\lim_{\sigma \to 0+} \frac{1}{\sigma} \psi_1(\sigma)\,\psi_2(\sigma) = 0 \tag{7.19}$$

by (7.15).

7.6. Lemma. *Let* $\sigma \in \mathbb{R}_0^+$. *Then*

$$\mathcal{A}(\sigma/2)\,2\,(1 + \psi_2(\sigma/2)) + 2\,\psi_1(\sigma/2)\,\psi_2(\sigma/2) \le \mathcal{A}(\sigma). \tag{7.20}$$

Proof. We have

$$2\,\Psi(\sigma/2) + 2\,\psi_1(\sigma/2)\,\psi_2(\sigma/2) = \Psi(\sigma) \quad \text{for } \sigma \in \mathbb{R}_0^+.$$

Hence

$$2\,\mathcal{A}\,(\sigma/2)\,\exp\Big(-\sum_{j=1}^{\infty}\psi_2(\sigma 2^{-j-1})\Big)+2\,\psi_1(\sigma/2)\,\psi_2(\sigma/2)$$

$$=\mathcal{A}\,(\sigma)\,\exp\Big(-\sum_{j=1}^{\infty}\psi_2(\sigma\,2^{-j})\Big)$$

and

$$2\,\mathcal{A}\,(\sigma/2)\,\exp\psi_2(\sigma/2)+2\,\psi_1(\sigma/2)\,\psi_2(\sigma/2)\,\exp\sum_{j=1}^{\infty}\psi_2(\sigma 2^{-j})=\mathcal{A}\,(\sigma)$$

for $\sigma\in\mathbb{R}_0^+$. Therefore (7.20) is correct. $\qquad\qquad\square$

7.7. Lemma. *There exists $\xi_1>0$ such that*

$$\mathcal{B}\,(\sigma)\leq\sigma\quad for\ \ 0\leq\sigma\leq\xi_1. \tag{7.21}$$

Proof.   (7.21) is a consequence of (7.18). $\qquad\qquad\square$

7.8. Lemma. *Let $0\leq\sigma\leq\xi_1$. Then*

$$\left.\begin{array}{l}\mathcal{B}^2(\sigma/2)+\mathcal{B}\,(\sigma/2)\,2\,(1+\psi_2\,(\sigma/2))\\[4pt]+\mathcal{A}\,(\sigma/2)\,\psi_1(\sigma/2)+3\,\psi_1(\sigma/2)\,\psi_2(\sigma/2)\leq\mathcal{B}\,(\sigma).\end{array}\right\} \tag{7.22}$$

Proof. Φ fulfils

$$2\,\Phi(\sigma/2)+\varphi(\sigma/2)=\Phi(\sigma)\quad for\ \ \sigma\in\mathbb{R}_0^+.$$

Hence

$$2\,\mathcal{B}\,(\sigma/2)\,\exp\Big(-\sum_{j=1}^{\infty}\psi_2(\sigma 2^{-j-1})-\sigma/2\Big)+\varphi(\sigma/2)$$

$$=\mathcal{B}\,(\sigma)\,\exp\Big(-\sum_{j=1}^{\infty}\psi_2(\sigma 2^{-j})-\sigma\Big),$$

$$2\,\mathcal{B}\,(\sigma/2)\,\exp\Big(\psi_2(\sigma/2)+\sigma/2\Big)+\varphi(\sigma/2)\,\exp\Big(\sum_{j=1}^{\infty}\psi_2(\sigma 2^{-j})+\sigma\Big)$$

$$=\mathcal{B}\,(\sigma),$$

$$\mathcal{B}\,(\sigma/2)\,2\,[1+\psi_2(\sigma/2)+\sigma/2]+\varphi(\sigma/2)\leq\mathcal{B}\,(\sigma)$$

for $\sigma\in\mathbb{R}_0^+$ and (7.22) is valid by virtue of Lemma 7.7 and (7.8). $\qquad\square$

Chapter 8

Strong Riemann solutions of generalized differential equations: a survey

A theory of SR-solutions of generalized differential equations is developed in Chapters 9–13. Its brief survey is presented below.

8.1. Notation. Let X_1, X_2 be vector spaces, $\Omega \subset X_1$ and let P be any set, $f : \Omega \to X_2$, $F : \Omega \times P \to X_2$. Denote

$$\left.\begin{aligned}
\Delta_v f(x) &= f(x+v) - f(x), \\
\Delta_v F(x,p) &= F(x+v,p) - F(x,p)
\end{aligned}\right\} \tag{8.1}$$

for x, $x+v \in \Omega$, $p \in P$.

Let X be a Banach space with the norm $\| \cdot \|$, $B(r) = \{x \in X; \|x\| \leq r\}$ for $0 \leq r < \infty$. Let $[a,b] \subset \mathbb{R}$, $R > 0$, $G : B(8R) \times [a,b]^2 \to X$.

8.2. Assumptions. Let G fulfil

$$G \quad \text{is continuous}, \tag{8.2}$$

$$\|G(x,\tau,t) - G(x,\tau,s)\| \leq \psi_2(t-s), \tag{8.3}$$

$$\|\Delta_v \left(G(x,\tau,t) - G(x,\tau,s)\right)\| \leq \|v\| \psi_1(t-s), \tag{8.4}$$

$$\|\Delta_w \Delta_v(G(x,\tau,t) - G(x,\tau,s))\| \leq \|w\| \, \|v\| \, \psi_1(t-s), \tag{8.5}$$

$$\left.\begin{aligned}
&\|G(x,\tau,t) - G(x,\tau,s) - G(x,\sigma,t) + G(x,\sigma,s)\| \\
&\qquad \leq \psi_1(t-s) \, \psi_2(\tau - \sigma),
\end{aligned}\right\} \tag{8.6}$$

$$\left.\begin{aligned}
&\|\Delta_v \left(G(x,\tau,t) - G(x,\tau,s) - G(x,\sigma,t) + G(x,\sigma,s)\right)\| \\
&\qquad \leq \|v\| \, \psi_1(t-s) \, \psi_2(\tau - \sigma)
\end{aligned}\right\} \tag{8.7}$$

for x, $x+v$, $x+w$, $x+v+w \in B(8R)$, t, s, τ, $\sigma \in [a,b]$, $s \leq t$, $\sigma \leq \tau$.

8.3. Proposition (Existence and uniqueness).
There exists $\xi_5 > 0$ having the following property:

If $b - a \leq \xi_5$, $s \in [a, b]$, $y \in B(3R)$, then there exists an SR-solution $u : [a, b] \to B(4R)$ of

$$\frac{\mathrm{d}}{\mathrm{d}t} x = D_t G(x, \tau, t) \tag{8.8}$$

fulfilling $u(s) = y$. Moreover, u is unique.

(See Theorems 12.10, 12.12.)

8.4. Proposition. *Let $b - a \leq \xi_5$ and let $u : [a, b] \to B(4R)$ be a solution of* (8.8). *Then*

$$\lim_{t \to \tau} (t - \tau)^{-1} \| u(t) - u(\tau) - G(u(\tau), \tau, t) + G(u(\tau), \tau, \tau) \| = 0. \tag{8.9}$$

(8.9) *implies that*

$$\frac{\partial}{\partial t} (u(t) - G(u(\tau), \tau, t))|_{t=\tau} = 0 \quad \text{for } \tau \in [a, b]. \tag{8.10}$$

E.g., if $G(x, \tau, \cdot)$ is nowhere differentiable then u is nowhere differentiable.

(See Lemma 12.7 and Corollary 12.8.)

8.5. Proposition (Continuous dependence).
There exists $\widehat{\Omega} : \mathbb{R}^+ \to \mathbb{R}^+$, $\widehat{\Omega}(\sigma) \to 0$ for $\sigma \to 0$ with the following property:

Let $G^* : B(8R) \times [a, b]^2 \to X$. Assume that $b - a \leq \xi_5$ and that G^* fulfils (8.2)–(8.7). *Put*

$$\left. \begin{array}{l} \text{dist}\,(G, G^*) \\ \quad = \sup \{ \| G(x, \tau, t) - G^*(x, \tau, t) \| \, ; x \in B(8R),\, \tau, t \in [a, b] \}\,. \end{array} \right\} \tag{8.11}$$

Let $y \in B(3R)$, $s \in [a, b]$. Let $v : [a, b] \to B(8R)$ be an SR-solution of (8.8), *$v(s) = y$, and let $v^* : [a, b] \to B(8R)$ be an SR-solution of*

$$\frac{\mathrm{d}}{\mathrm{d}\tau} x = D_t G^*(x, \tau, t), \quad v^*(s) = y\,. \tag{8.12}$$

Then

$$v(t),\, v^*(t) \in B(4R) \quad \text{for } t \in [a, b] \tag{8.13}$$

and

$$\|v(t) - v^*(t)\| \leq \widehat{\Omega}(\text{dist}(G, G^*)) \quad for \ \ t \in [a, b]. \tag{8.14}$$

Proof. (8.13) follows from Proposition 8.3 and (8.14) is a consequence of Theorem 13.7 where $\widehat{\Omega} = \frac{3}{2}\varkappa\Omega$. $\square$

8.6. Remark. Let e.g. $[a, b] \subset \mathbb{R}$, $\frac{1}{2} < \alpha \leq 1$, $0 < \beta \leq 1$, $\alpha + \beta > 1$, $\widehat{R} > 0$,

$$G(x, \tau, t)$$
$$= -h(x, \tau + \varepsilon^\alpha \sin(\tau/\varepsilon))\,\varepsilon^\alpha \sin(t/\varepsilon) - \lambda\,p\,(\tau + \varepsilon^\alpha \sin(\tau/\varepsilon))\,\varepsilon^\beta \sin\,(t/(\lambda\varepsilon)),$$

where $0 < \varepsilon \leq 1$, $0 \leq \lambda \leq 1$,

$$h : B(\widehat{R}) \times [a - 1, b + 1] \to X, \quad p : [a - 1, b + 1] \to X,$$
$$D_1 h(x, \tau) = \frac{\partial}{\partial x}\,h(x, \tau) \quad \text{is Lipschitzian with respect to} \ \ x, \tau$$

and

$$D p\,(s) = \frac{\mathrm{d}}{\mathrm{d}s}\,p(s) \quad \text{is Lipschitzian}.$$

Then any solution of the classical equation

$$\dot{x} = h(x, t + \varepsilon^\alpha \,\sin(t/\varepsilon))\,\varepsilon^{\alpha-1}\,\cos(t/\varepsilon) + p(t + \varepsilon^\alpha \,\sin(t/\varepsilon))\,\varepsilon^{\beta-1}\,\cos(\lambda t/\varepsilon)$$

is a solution of (8.8) and vice versa.

Conditions (7.1)–(7.4) and (8.2)–(8.7) are fulfilled if

$$\psi_1(\sigma) = \varkappa\sigma^\alpha, \quad \psi_2(\sigma) = \varkappa\sigma^\beta \quad \text{for } 0 \leq \sigma \leq 1$$

and

$$\psi_1(\sigma) = \psi_2(\sigma) = \varkappa\sigma \quad \text{for } 1 \leq \sigma.$$

The existence and uniqueness of solutions of (8.8) are given by Proposition 8.4. Observe that β may be small if α is sufficiently close to 1.

Chapter 9

Approximate solutions: boundedness

9.1 . Notation. Let $[a, b] \subset \mathbb{R}$, $c = a + 2(b - a)$, $\xi_2 \in \mathbb{R}^+$, $b - a \le \xi_2$, $G : B(8R) \times [a, b]^2 \to X$. Assume that G fulfils (8.2)– (8.7). Further, assume that

$$\psi_2(2\xi_2) + \mathcal{A}(2\xi_2) < R, \quad b - a \le \xi_2, \tag{9.1}$$

where the functions ψ_1 and $\mathcal{A}$ were introduced in Notation 7.1.

Extend G by putting

$$\widetilde{G}(x, \tau, t) = G(x, \min\{\tau, b\}, \min\{t, b\}) \text{ for } (\tau, t) \in [a, c]^2, \ x \in B(8R) \tag{9.2}$$

and for S, t, T such that $S, t, T \in [a, c]$ and $S < T$ let $\mathbf{G}(S, t, T)$ be a mapping from $B(8R)$ to X, the value of $\mathbf{G}(S, t, T)$ at x being denoted by $x\,\mathbf{G}(S, t, T)$ and

$$x\,\mathbf{G}(S, t, T) = \begin{cases} 0 & \text{if } t \le S, \\ \widetilde{G}(x, S, t) - \widetilde{G}(x, S, S) & \text{if } S \le t \le T, \\ \widetilde{G}(x, S, T) - \widetilde{G}(x, S, S) & \text{if } T \le t. \end{cases} \tag{9.3}$$

9.2 . Remark. Existence, uniqueness and continuous dependence of solutions of (8.8) will be proved for $t \in [a, b]$. The proofs are based on the properties of the functions in the form

$$\left. \begin{array}{l} x\,(id + \mathbf{G}(S, t, S + \sigma))\,(id + \mathbf{G}(S + \sigma, t, S + 2\,\sigma)) \\[4pt] \dots (id + \mathbf{G}(S + (k-1)\,\sigma, t, S + k\,\sigma)) \quad \text{for } k \in \mathbb{N}, \end{array} \right\} \tag{9.4}$$

where

$$x\,(id + \mathbf{G}(S, t, S + \sigma)) = x + x\,\mathbf{G}\,(S, t, S + \sigma).$$

In Chapter 12 functions in the form (9.4) are needed such that $S \in [a, b]$, $\sigma = (b - a)\,2^{-j}$, $k = 1, 2, \dots, 2^j$, $j \in \mathbb{N}$, so that $S + 2^j\,\sigma = S + b - a$ and

$[S, S+b-a] \subset [a,c]$ (see (12.3)). This is the motivation for extending G to $\widetilde{G}$.

9.3. Lemma. $\widetilde{G}$ *fulfils (8.2)–(8.7) on its domain*

$$B(8R) \times \{(S,t,T); [S,T] \subset [a,c],\ t \in [a,c]\}. \tag{9.5}$$

Moreover,

$$\mathbf{G} \text{ is continuous}, \tag{9.6}$$

$$\|x\,\mathbf{G}(S,t,T)\| \le \psi_2(t-S), \tag{9.7}$$

$$\|\Delta_v(x\,\mathbf{G}(S,t,T))\| \le \|v\|\,\psi_1(t-S), \tag{9.8}$$

$$\|\Delta_w\,\Delta_v(x\mathbf{G}(S,t,T))\| \le \|w\|\,\|v\|\,\psi_1(t-S), \tag{9.9}$$

$$\left.\begin{array}{l}\left\|x\,\mathbf{G}(S,t,\tfrac{S+T}{2}) + x\,\mathbf{G}(\tfrac{S+T}{2},t,T) - x\,\mathbf{G}(S,t,T)\right\| \\ \le \psi_1\left(\tfrac{t-S}{2}\right)\psi_2\left(\tfrac{t-S}{2}\right),\end{array}\right\} \tag{9.10}$$

$$\left.\begin{array}{l}\left\|\Delta_v\left(x\,\mathbf{G}(S,t,\tfrac{S+T}{2}) + x\,\mathbf{G}(\tfrac{S+T}{2},t,T) - x\,\mathbf{G}(S,t,T)\right)\right\| \\ \le \|v\|\,\psi_1(\tfrac{T-S}{2})\,\psi_2(\tfrac{T-S}{2})\end{array}\right\} \tag{9.11}$$

for x, $x+v$, $x+v+w$, $x+w \in B(8R)$, $t \in [S,T] \subset [a,c]$.

Proof. $\widetilde{G}$ fulfils (8.2)–(8.7) on its domain by (9.2) since ψ_1, ψ_2 are non-decreasing.

Relation (9.6) is a consequence of (8.2) and (9.7), (9.8), (9.9) follow directly from (8.3), (8.4), (8.5).

If $t \le \tfrac{S+T}{2}$ then the left hand side of (9.10) vanishes.

Let $\tfrac{S+T}{2} < t \le T$. Then

$$x\,\mathbf{G}(S,t,\tfrac{S+T}{2}) + x\,\mathbf{G}(\tfrac{S+T}{2},t,T) - x\,\mathbf{G}(S,t,T)$$
$$= \widetilde{G}(x,S,\tfrac{S+T}{2}) - \widetilde{G}(x,S,S) + \widetilde{G}(x,\tfrac{S+T}{2},t)$$
$$- \widetilde{G}(x,\tfrac{S+T}{2},\tfrac{S+T}{2}) - \widetilde{G}(x,S,t) + \widetilde{G}(x,S,S)$$

and (9.10) holds by (8.6). (9.11) is proved in an analogous way (cf. (8.7)). $\qquad\sqcap$

9.4. Definition. Let

$$x \in B(8R), \quad [S,T] \subset [a,c], \quad t \in [a,c], \quad i \in \mathbb{N}_0, \quad j \in \{0,1,\ldots,2^i\}.$$

Put

$$x\,(id + \mathbf{G}(S,t,T)) = x + x\,\mathbf{G}(S,t,T),$$
$$Z(i,j) = S + j\,\frac{T-S}{2^i}\,.$$

Assume that

$$\left.\begin{array}{l}
x\,(id + \mathbf{G}(Z(i,0), Z(i,1), Z(i,1))) \\[4pt]
\quad (id + \mathbf{G}(Z(i,1), Z(i,2), Z(i,2))) \ldots \\[4pt]
\quad\quad (id + \mathbf{G}(Z(i,j-2), Z(i,j-1), Z(i,j-1))) \\[4pt]
\quad\quad\quad (id + \mathbf{G}(Z(i,j-1), t, Z(i,j))) \in B(8R)
\end{array}\right\} \tag{9.12}$$

for $j = 1, 2, \ldots, 2^i$ and $Z(i, j-1) < t \le Z(i,j)$. Then denote by $x\,V_i(S,t,T)$ the expression in (9.12), $S < t \le T$ and put

$$x\,V_i(S,t,T) = x \qquad\qquad \text{for } a \le t \le S, \tag{9.13}$$

$$x\,V_i(S,t,T) = x\,V_i(S,T,T) \quad \text{for } T < t \le c \tag{9.14}$$

if $x\,V_i(S,T,T)$ exists.

9.5. Remark. Let $u_i(t) = x\,V_i(S,t,T)$ for $S \le t \le T$. u_i is called an *approximate solution* of (8.8), which is justified by (11.16), (12.3), Theorem 12.10.

9.6. Lemma. *Let $t \in [a,c]$, $i \in \mathbb{N}_0$, $x \in B(8R)$ and let $x\,V_i(S,T,T)$ exist. Then*

$$\left.\begin{array}{l}
x\,V_i(S,t,T) \\[4pt]
= x\,(id + \mathbf{G}(S,t,Z(i,1)))\,(id + \mathbf{G}(Z(i,1),t,Z(i,2))) \\[4pt]
\quad \ldots (id + \mathbf{G}(Z(i,2^i-1),t,T)) \quad \text{for } t \in [a,c]\,.
\end{array}\right\} \tag{9.15}$$

Proof. (9.15) is a consequence of (9.3) and Definition 9.4. $\qquad\qquad \square$

9.7. Lemma. *Let $i \in \mathbb{N}$, $t \in [S,T] \subset [a,c]$, $x \in B(8R)$ and let $x\,V_i(S,T,T)$ exist. Then*

$$x\,V_i(S,t,T) = x\,V_{i-1}(S,t,\tfrac{S+T}{2})\,V_{i-1}(\tfrac{S+T}{2},t,T) \tag{9.16}$$

and

$$x\,V_{i-1}(S,t,\tfrac{S+T}{2}) = x\,V_i(S,t,T) \quad \textit{if } t \le \tfrac{S+T}{2} \tag{9.17}$$

Proof. (9.17) is a consequence of Definition 9.4. If $t \le \tfrac{S+T}{2}$ then (9.16) is valid since

$$x\,V_{i-1}(S,t,\tfrac{S+T}{2})\,V_{i-1}(\tfrac{S+T}{2},t,T) = z\,V_{i-1}(S,t,\tfrac{S+T}{2})$$

by (9.17). If $t > \tfrac{S+T}{2}$ then (9.16) follows by Definition 9.4.   □

9.8. Definition. For $[S,T] \subset [a,c]$, $t \in [a,c]$, $x \in B(8R)$ and $i \in \mathbb{N}_0$, put

$$x\,Y_i(S,t,T) = x\,V_i(S,t,T) - x - x\,\mathbf{G}(S,t,T) \tag{9.18}$$

if $x\,V_i(S,T,T)$ exists.

9.9. Lemma. *Let $i \in \mathbb{N}_0$, $t \in [S,T] \subset [a,c]$, $x \in B(8R)$ and let $x\,V_{i+1}(S,T,T)$ exist. Then*

$$\left.\begin{aligned}
x\,Y_{i+1}(S,t,T) &= x\,V_i(S,t,\tfrac{S+T}{2})Y_i(\tfrac{S+T}{2},t,T) + x\,Y_i(S,t,\tfrac{S+T}{2}) \\
&+ \left(x\,Y_i(S,t,\tfrac{S+T}{2}) + x + x\,\mathbf{G}(S,t,\tfrac{S+T}{2})\right)\mathbf{G}(\tfrac{S+T}{2},t,T) \\
&+ x\,\mathbf{G}(S,t,\tfrac{S+T}{2}) - x\,\mathbf{G}(S,t,T)
\end{aligned}\right\} \tag{9.19}$$

and

$$\left.\begin{aligned}
\|x\,Y_{i+1}(S,t,T)\| &\le \left\|x\,V_i(S,t,\tfrac{S+T}{2})\,Y_i(\tfrac{S+T}{2},t,T)\right\| \\
&+ \left\|x\,Y_i\left(S,t,\tfrac{S+T}{2}\right)\right\|\left(1+\psi_1\left(\tfrac{T-S}{2}\right)\right) + 2\,\psi_1(\tfrac{T-S}{2})\,\psi_2(\tfrac{T-S}{2}).
\end{aligned}\right\} \tag{9.20}$$

Proof. By (9.16) and by Definitions 9.4, 9.8, we have

$$\begin{aligned}
x\,Y_{i+1}&(S,t,T) \\
&= x\,V_i(S,t,\tfrac{S+T}{2})\left[Y_i(\tfrac{S+T}{2},t,T) + id + \mathbf{G}(\tfrac{S+T}{2},t,T)\right] \\
&\quad - x - x\,\mathbf{G}(S,t,T) \\
&= x\,V_i(S,t,\tfrac{S+T}{2})\,Y_i(\tfrac{S+T}{2},t,T) \\
&\quad + x\,Y_i(S,t,\tfrac{S+T}{2}) + x + x\,\mathbf{G}(S,t,\tfrac{S+T}{2}) \\
&\quad + \left(x\,Y_i(S,t,\tfrac{S+T}{2}) + x + x\,\mathbf{G}(S,t,\tfrac{S+T}{2})\right)\mathbf{G}(\tfrac{S+T}{2},t,T) \\
&\quad - x - x\,\mathbf{G}(S,t,T)
\end{aligned}$$

since $x\,V_i(S,t,\frac{S+T}{2})\in B(8R)$ and $x\,V_i(S,t,\frac{S+T}{2})\,\mathbf{G}(\frac{S+T}{2},t,T)$ exists. Therefore

$$
\left.
\begin{aligned}
&x\,Y_{i+1}(S,t,T)\\
&= x\,V_i(S,t,\tfrac{S+T}{2})\,Y_i(\tfrac{S+T}{2},t,T)\\
&\quad + x\,Y_i(S,t,\tfrac{S+T}{2})+x+x\,\mathbf{G}(S,t,\tfrac{S+T}{2})\\
&\quad + \left[x\,Y_i(S,t,\tfrac{S+T}{2})+x+x\,\mathbf{G}(S,t,\tfrac{S+T}{2})\right]\mathbf{G}(\tfrac{S+T}{2},t,T)\\
&\quad - x - x\,\mathbf{G}(S,t,T)\,,
\end{aligned}
\right\}
\tag{9.21}
$$

Hence (9.19) holds. Moreover, (9.21) can be written in the form

$$
\begin{aligned}
&x\,Y_{i+1}(S,t,T)\\
&= x\,V_i(S,t,\tfrac{S+T}{2})\,Y_i(\tfrac{S+T}{2},t,T)+x\,Y_i(S,t,\tfrac{S+T}{2})\\
&\quad + \left[\left(x\,Y_i(S,t,\tfrac{S+T}{2})+x+x\,\mathbf{G}(S,t,\tfrac{S+T}{2})\right)\mathbf{G}(\tfrac{S+T}{2},t,T)\right.\\
&\qquad \left. - \left(x+x\,\mathbf{G}(S,t,\tfrac{S+T}{2})\right)\mathbf{G}(\tfrac{S+T}{2},t,T)\right]\\
&\quad + \left[\left(x+x\,\mathbf{G}(S,t,\tfrac{S+T}{2})\right)\mathbf{G}(\tfrac{S+T}{2},t,T)-x\,\mathbf{G}(\tfrac{S+T}{2},t,T)\right]\\
&\quad + \left[x\,\mathbf{G}(S,t,\tfrac{S+T}{2})+x\,\mathbf{G}(\tfrac{S+T}{2},t,T)-x\,\mathbf{G}(S,t,T)\right].
\end{aligned}
$$

The first bracket does not exceed $\|x\,Y_i(S,t,\frac{S+T}{2})\|\,\psi_1(\frac{T-S}{2})$ by (9.8), the second one and the third one are estimated by $\psi_1(\frac{T-S}{2})\,\psi_2(\frac{T-S}{2})$. Therefore (9.20) is correct. $\qquad\square$

9.10. Lemma. *Let* $i\in\mathbb{N}_0$, $t\in[S,T]\subset[a,c]$, $x\in B(8R)$ *and let* $x\,V_i(S,T,T)$ *exist. Then*

$$
\|x\,Y_i(S,t,T)\|\le \mathcal{A}(T-S)
\tag{9.22}
$$

$$
\|x\,V_i(S,t,T)-x\|\le \mathcal{A}(T-S)+\psi_2(T-S)
\tag{9.23}
$$

Proof. PART 1. By Definitions 9.4, 9.8

$$
x\,V_0(S,t,T)=x\,\mathbf{G}(S,t,T)\,,
$$

and

$$
x\,Y_0(S,t,T)=0\,.
\tag{9.24}
$$

Lemma 9.10 is valid for $i = 0$ (cf. (9.7)).

PART 2. Assume that there exists $k \in \mathbb{N}$ such that Lemma 9.10 is valid for $i = 0, 1, \dots, k-1$.

Let $t \in [S, T] \subset [a, c]$, $x \in B(8R)$ and let $x V_k(S, T, T)$ exist. By (9.20)

$$\left.\begin{aligned}
\|x Y_k(S, t, T)\| &\leq \|x V_{k-1}(S, t, \tfrac{S+T}{2}) Y_{k-1}(\tfrac{S+T}{2}, t, T)\| \\
&+ \|x Y_{k-1}(S, t, \tfrac{S+T}{2})\| \left(1 + \psi_1(\tfrac{T-S}{2})\right) + 2\,\psi_1(\tfrac{T-S}{2})\,\psi_2(\tfrac{T-S}{2}).
\end{aligned}\right\} \quad (9.25)$$

If $t \leq \tfrac{S+T}{2}$ then (cf. (9.18), (9.3))

$$x V_{k-1}(S, t, \tfrac{S+T}{2}) Y_{k-1}(\tfrac{S+T}{2}, t, T) = 0, \quad \|x Y_{k-1}(S, t, \tfrac{S+T}{2})\| \leq \mathcal{A}(\tfrac{T-S}{2})$$

and (9.22) holds by (7.20). If $t > \tfrac{S+T}{2}$ then

$$\|x V_{k-1}(S, t, \tfrac{S+T}{2}) Y_{k-1}(\tfrac{S+T}{2}, t, T)\| \leq \mathcal{A}(\tfrac{T-S}{2}),$$

$$\|x Y_{k-1}(S, t, \tfrac{S+T}{2})\| \leq \mathcal{A}(\tfrac{T-S}{2})$$

and (9.22) holds by (7.20).

PART 3. Parts 1 and 2 imply that (9.22) holds. Moreover, (9.23) follows by (9.22), (9.18), (9.8). $\qquad\square$

9.11. Theorem. *Let* $i \in \mathbb{N}_0$, $[S, T] \subset [a, c]$, $t \in [a, c]$ *and* $x \in B(7R)$. *Then*

$$\|x V_i(S, t, T) - x\| < R. \qquad (9.26)$$

Proof. PART 1. By Definition 9.4, (9.7), (9.1),

$$x V_0(S, t, T) = x + x\,\mathbf{G}(S, t, T),$$

$$\|x V_0(S, t, T) - x\| \leq \|x\,\mathbf{G}(S, t, T)\| \leq \psi_2(t-S) < R$$

if $S \leq t \leq T$. Moreover,

$$x V_0(S, t, T) = x \qquad \text{for } a \leq t < T,$$

$$x V_0(S, t, T) = x V_0(S, T, T) \quad \text{for } T < t \leq C.$$

Hence Theorem 9.11 is valid if $i = 0$.

PART 2. Assume that there exists $k \in \mathbb{N}$ such that (9.26) holds for $i = 0, 1, \dots, k-1$, $S \leq t \leq T$ and that (9.26) does not hold for $i = k$ and all

$x \in B(7R)$ and $t \in [S, T] \subset [a, c]$. By (9.6) and Definition 9.4, $\mathbf{G}(S, ., T)$ is continuous. Hence there exist $[\widehat{S}, \widehat{T}] \subset [a, c]$ and $\widehat{x} \in B(7R)$ such that

$$\|\widehat{x} V_k(\widehat{S}, t, \widehat{T}) - \widehat{x}\| < R \quad \text{if} \quad \widehat{S} \leq t < T$$

and

$$\|\widehat{x} V_k(\widehat{S}, \widehat{T}, \widehat{T}) - \widehat{x}\| = R \tag{9.27}$$

By Lemma 9.10 and (9.1)

$$\|\widehat{x} V_k(\widehat{S}, \widehat{T}, \widehat{T}) - \widehat{x}\| < \mathcal{A}(\widehat{T} - \widehat{S}) + \psi_2(\widehat{T} - \widehat{S}) < R$$

which contradicts (9.27). Therefore

$$\|x V_k(S, t, T) - x\| < R \quad \text{for} \quad t \in [S, T] \subset [a, c], \ x \in B(7R).$$

By Definition 9.4

$$\|x V_k(S, t, T) - x\| < R \quad \text{for} \quad [S, T] \subset [a, c], \ t \in [a, c], \ x \in B(7R).$$

PART 3. Parts 1 and 2 imply that Theorem 9.11 is valid. $\qquad\square$

Chapter 10

Approximate solutions: a Lipschitz condition

10.1. Notation. Observe that $\displaystyle\sum_{j=1}^{\infty} \psi_1(\sigma\, 2^{-j}) \to 0$ for $\sigma \to 0$ (cf. (7.1), (7.2), (7.4)). Let $\xi_3 \in \mathbb{R}^+$ fulfil (cf. (7.21), (9.1), (7.1), (7.2), (7.4))

$$\xi_3 \leq \min\{\xi_1, \xi_2\} \tag{10.1}$$

(ξ_1 having been introduced by Lemma 7.7, ξ_2 in (9.1)),

$$\exp\Big(2\,\xi_3 + \sum_{j=1}^{\infty} \psi_1(2\,\xi_3\, 2^{-j})\Big) \leq 2\,, \tag{10.2}$$

$$\mathcal{A}(2\,\xi_3) + \psi_2(2\,\xi_3) \leq R\,. \tag{10.3}$$

Assume that G fulfils (8.2)–(8.7).

10.2. Theorem. *Let*

$$[a, c] \subset \mathbb{R}, \ c-a \leq 2\,\xi_3\,, \tag{10.4}$$

$$[S, T] \subset [a, c], \ i \in \mathbb{N}_0, \ x, z \in B(6R), \ t \in [a, c]\,. \tag{10.5}$$

Then

$$\left.\begin{aligned}
\| x\, V_i(S, t, T) &- z\, V_i(S, t, T) - x + z \\
&\qquad - x\, \mathbf{G}(S, t, T) + z\, \mathbf{G}(S, t, T) \| \\
&= \| x\, Y_i(S, t, T) - z\, Y_i(S, t, T) \| \\
&\leq \| x - z \|\, \mathcal{B}(T - S)\,.
\end{aligned}\right\} \tag{10.6}$$

The proof is postponed after Lemma 10.3.

10.3. Lemma. *Assume that there exists $\ell \in \mathbb{N}_0$ such that*

$$\|\bar{x}\, Y_\ell(\bar{S}, \bar{t}, \bar{T}) - \bar{z}\, Y_\ell(\bar{S}, \bar{t}, \bar{T})\| \le \|\bar{x} - \bar{z}\| \, \mathcal{B}(\bar{T} - \bar{S}) \tag{10.7}$$

for

$$[\bar{S}, \bar{T}] \subset [a, c], \quad \bar{x}, \bar{z} \in B(6R), \ \bar{t} \in [a, c]. \tag{10.8}$$

Then

$$\left.\begin{aligned} \|x\, Y_{\ell+1}(S, t, T) - z\, Y_{\ell+1}(S, t, T)\| \le \|x - z\| \, \mathcal{B}(T - S) \\ \text{for } [S, T] \subset [a, c], \ x, z \in B(6R), \ , \ t \in [a, c]. \end{aligned}\right\} \tag{10.9}$$

Proof.   Let

$$[S, T] \subset [a, c], \ x, z \in B(6R), \, , \ t \in [a, c].$$

Observe that (cf. (9.22), (9.6))

$$\left.\begin{aligned} \|x\, Y_\ell(S, t, \tfrac{S+T}{2})\| \le \mathcal{A}(\tfrac{T-S}{2}), \\ \|x\, \mathbf{G}(S, t, \tfrac{S+T}{2})\| \le \psi_2(\tfrac{T-S}{2}) \end{aligned}\right\} \tag{10.10}$$

and that (cf. (10.3))

$$x\, Y_\ell(S, t, \tfrac{S+T}{2}) + z + x\, \mathbf{G}(S, t, \tfrac{S+T}{2}) \in B(8R) \tag{10.11}$$

since $\|z\| \le 6R$. Write ℓ instead of i in (9.19), then write z instead of x and form the difference of these equations. In this way we obtain

$$x\, Y_{\ell+1}(S, t, T) - z\, Y_{\ell+1}(S, t, T) = \mathcal{F}_1 + \mathcal{F}_2 + \mathcal{F}_3 + \mathcal{F}_4 + \mathcal{F}_5, \tag{10.12}$$

where

$$\left.\begin{aligned} \mathcal{F}_1 = x\, V_\ell(S, t, \tfrac{S+T}{2})\, Y_\ell(\tfrac{S+T}{2}, t, T) \\ - z\, V_\ell(S, t, \tfrac{S+T}{2})\, Y_\ell(\tfrac{S+T}{2}, t, T), \end{aligned}\right\} \tag{10.13}$$

$$\mathcal{F}_2 = x\, Y_\ell(S, t, \tfrac{S+T}{2}) - z\, Y_\ell(S, t, \tfrac{S+T}{2}), \tag{10.14}$$

$$\left.\begin{aligned} \mathcal{F}_3 = [x\, Y_\ell(S, t, \tfrac{S+T}{2}) + x + x\, \mathbf{G}(S, t, \tfrac{S+T}{2})]\, \mathbf{G}(\tfrac{S+T}{2}, t, T) \\ - x\, \mathbf{G}(\tfrac{S+T}{2}, t, T) + z\, \mathbf{G}(\tfrac{S+T}{2}, t, T) \\ -[x\, Y_\ell(S, t, \tfrac{S+T}{2}) + z + x\, \mathbf{G}(S, t, \tfrac{S+T}{2})]\, \mathbf{G}(\tfrac{S+T}{2}, t, T), \end{aligned}\right\} \tag{10.15}$$

$$\left.\begin{aligned}
\mathcal{F}_4 &= [x\, Y_\ell(S,t,\tfrac{S+T}{2}) + z + x\, \mathbf{G}(S,t,\tfrac{S+T}{2})]\, \mathbf{G}(\tfrac{S+T}{2},t,T) \\
&\quad - [z\, Y_\ell(S,t,\tfrac{S+T}{2}) + z + z\, \mathbf{G}(S,t,\tfrac{S+T}{2})]\, \mathbf{G}(\tfrac{S+T}{2},t,T),
\end{aligned}\right\} \tag{10.16}$$

$$\left.\begin{aligned}
\mathcal{F}_5 &= x\, \mathbf{G}(S,t,\tfrac{S+T}{2}) + x\, \mathbf{G}(\tfrac{S+T}{2},t,T) - x\, \mathbf{G}(S,t,T) \\
&\quad - z\, \mathbf{G}(S,t,\tfrac{S+T}{2}) - z\, \mathbf{G}(\tfrac{S+T}{2},t,T) + z\, \mathbf{G}(S,t,T).
\end{aligned}\right\} \tag{10.17}$$

The following estimates are deduced from (10.13)–(10.17) by (10.7), (9.7)–(9.11), (9.23).

In particular, by (10.7), (9.18) and (9.8), we have

$$\left.\begin{aligned}
\|\mathcal{F}_1\| &\le \|x\, V_\ell(S,t,\tfrac{S+T}{2}) - z\, V_\ell(S,t,\tfrac{S+T}{2})\|\, \mathcal{B}(\tfrac{T-S}{2}) \\
&\le \|x - z\| \left(1 + \psi_2(\tfrac{T-S}{2}) + \mathcal{B}(\tfrac{T-S}{2})\right) \mathcal{B}(\tfrac{T-S}{2}).
\end{aligned}\right\} \tag{10.18}$$

Further, by (10.7), we have

$$\|\mathcal{F}_2\| \le \|x - z\|\, \mathcal{B}(\tfrac{T-S}{2}). \tag{10.19}$$

By (9.5), (9.9), (9.21)

$$\left.\begin{aligned}
\|\mathcal{F}_3\| &\le \|x-z\| \left\|x Y_\ell(S,t,\tfrac{S+T}{2}) + x\mathbf{G}(S,t,\tfrac{S+T}{2})\right\| \psi_1(\tfrac{T-S}{2}) \\
&\le \|x-z\| \left(\mathcal{A}(\tfrac{T-S}{2}) + \psi_2(\tfrac{T-S}{2})\right) \psi_1(\tfrac{T-S}{2})
\end{aligned}\right\} \tag{10.20}$$

since $\mathcal{F}_3 = -\,\Delta_w \Delta_v (x\, \mathbf{G}(\tfrac{S+T}{2},t,T)$ where

$$v = z - x, \quad w = x\, Y_\ell(S,t,\tfrac{S+T}{2}) + x\, \mathbf{G}(\tfrac{S+T}{2},t,T).$$

By (9.8), (10.6)

$$\left.\begin{aligned}
\|\mathcal{F}_4\| &\le \left\|x\, Y_\ell(S,t,\tfrac{S+T}{2}) - z\, Y_\ell(S,t,\tfrac{S+T}{2})\right. \\
&\quad \left. + x\, \mathbf{G}(S,t,\tfrac{S+T}{2}) - z\, \mathbf{G}(S,t,\tfrac{S+T}{2})\right\| \psi_1(\tfrac{T-S}{2}) \\
&\le \|x - z\| \left(\mathcal{B}(\tfrac{T-S}{2}) + \psi_2(\tfrac{T-S}{2})\right) \psi_1(\tfrac{T-S}{2}).
\end{aligned}\right\} \tag{10.21}$$

Finally, by (9.11)

$$\|\mathcal{F}_5\| \le \|x - z\|\, \psi_1(\tfrac{T-S}{2})\, \psi_2(\tfrac{T-S}{2}). \tag{10.22}$$

Now, (10.12) and (10.18)– (10.22) imply that

$$
\left.\begin{aligned}
&\|x\,Y_{\ell+1}(S,t,T) - z\,Y_{\ell+1}(S,t,T)\| \\
&\quad \leq \|x-z\|\,\mathcal{B}(\tfrac{T-S}{2})\left(2+\psi_1(\tfrac{T-S}{2})+\psi_2(\tfrac{T-S}{2})\right)+\mathcal{B}^2(\tfrac{T-S}{2}) \\
&\quad +3\,\psi_1(\tfrac{T-S}{2})\,\psi_2(\tfrac{T-S}{2})+\mathcal{A}(\tfrac{T-S}{2})\,\psi_1(\tfrac{T-S}{2})).
\end{aligned}\right\} \quad (10.23)
$$

(10.9) holds by (10.23), (7.2) and (7.22).   □

Proof of Theorem 10.2. (10.7) holds by (9.24) if $\ell = 0$. Lemma 10.3 implies that (10.6) holds if $\ell \in \mathbb{N}_0$.   □

Chapter 11

Approximate solutions: convergence

11.1. Notation. Let

$$
\left.
\begin{aligned}
&0 < \xi_4 \le \xi_3\,, \\[4pt]
&\psi_2(\xi_4) + \psi_1(\tfrac{\xi_4}{2})\,\psi_2(\tfrac{\xi_4}{2})\,\exp\Big(\tfrac{1}{2}\sum_{j=2}^{\infty} \mathcal{C}_j(\xi_4)\Big) < R \\[4pt]
&\text{where } \mathcal{C}_j(z) = \psi_2(\tfrac{z}{2^j}) + \mathcal{B}(\tfrac{z}{2^j}) \text{ for } z > 0,\, j \in \mathbb{N} \\[4pt]
&\text{and } \mathcal{B} \text{ has been introduced in } (7.9)\,,
\end{aligned}
\right\}
\tag{11.1}
$$

$$
a < c \le a + \xi_4\,.
\tag{11.2}
$$

11.2. Lemma. *Let*

$$
[S,T] \subset [a,c]\,, \qquad x \in B(7R)\,.
\tag{11.3}
$$

Then

$$
\| x\,V_1(S,t,T) - x\,V_0(S,t,T)\| \le 2\,\psi_1(\tfrac{T-S}{2})\,\psi_2(\tfrac{T-S}{2})\,.
\tag{11.4}
$$

Proof. Let (11.3) be fulfilled. By Definition 9.4

$$
\begin{aligned}
x\,V_0(S,t,T) &= x + x\,\mathbf{G}(S,t,T) && \text{for } S \le t \le T\,, \\
x\,V_1(S,t,T) &= x + x\,\mathbf{G}(S,t,\tfrac{S+T}{2}) && \text{for } S \le t \le \tfrac{S+T}{2}\,.
\end{aligned}
$$

Hence

$$
x\,V_1(S,t,T) - x\,V_0(S,t,T) = 0 \quad \text{for } t \le \tfrac{S+T}{2}\,.
$$

Let $\tfrac{S+T}{2} < t$. Then (cf. also (9.17), (9.16))

$$
\begin{aligned}
&x\,V_1(S,t,T) - x\,V_0(S,t,T) \\
&\qquad = x\,V_0(S,\tfrac{S+T}{2},\tfrac{S+T}{2})\,V_0(\tfrac{S+T}{2},t,T) - x\,V_0(S,t,T)
\end{aligned}
$$

$$= \left(x + x\,\mathbf{G}(S, \tfrac{S+T}{2}, \tfrac{S+T}{2})\right)\left(id + \mathbf{G}(\tfrac{S+T}{2}, t, T)\right)$$

$$- x - x\,\mathbf{G}(S, t, T)$$

$$= x + x\,\mathbf{G}(S, \tfrac{S+T}{2}, \tfrac{S+T}{2})$$

$$+ \left(x + x\,\mathbf{G}(S, \tfrac{S+T}{2}, \tfrac{S+T}{2})\right)\mathbf{G}(\tfrac{S+T}{2}, t, T)$$

$$- x\,\mathbf{G}(\tfrac{S+T}{2}, t, T) + x\,\mathbf{G}(\tfrac{S+T}{2}, t, T) - x - x\,\mathbf{G}(S, t, T)$$

and (11.4) holds since (cf. (9.7), (9.8), (9.10))

$$\left\| \left(x + x\,\mathbf{G}(S, \tfrac{S+T}{2}, \tfrac{S+T}{2})\right)\mathbf{G}(\tfrac{S+T}{2}, t, T) - x\,\mathbf{G}(\tfrac{S+T}{2}, t, T) \right\|$$

$$\le \psi_1(\tfrac{S+T}{2})\,\psi_2(\tfrac{S+T}{2}),$$

$$\left\| x\,\mathbf{G}(S, \tfrac{S+T}{2}, \tfrac{S+T}{2}) + x\,\mathbf{G}(\tfrac{S+T}{2}, t, T) - x\,\mathbf{G}(S, t, T) \right\|$$

$$\le \psi_1(\tfrac{T-S}{2})\,\psi_2(\tfrac{T-S}{2}).$$

Lemma 11.2 is correct. $\qquad\qquad\square$

11.3. Lemma. *Let* $i \in \mathbb{N}$. *Assume that*

$$\left.\begin{aligned}
&\|\widehat{x}\,V_i(\widehat{S}, \widehat{t}, \widehat{T}) - \widehat{x}\,V_{i-1}(\widehat{S}, \widehat{t}, \widehat{T})\| \\
&\qquad \le 2^i\,\psi_1(\tfrac{\widehat{T}-\widehat{S}}{2^i})\,\psi_2(\tfrac{\widehat{T}-\widehat{S}}{2^i})\exp\left(\tfrac{1}{2}\sum_{j=1}^{i-1}\mathcal{C}_j(\widehat{T}-\widehat{S})\right)
\end{aligned}\right\} \qquad (11.5)$$

for

$$[\widehat{S}, \widehat{T}] \subset [a, c], \quad \widehat{x} \in B(6R). \qquad (11.6)$$

(*Observe that for* $i = 1$ *we have* $\exp\left(\tfrac{1}{2}\overset{\circ}{\underset{j=1}{\sum}}\mathcal{C}_j(\widehat{T}-\widehat{S})\right) = 1$.)

Then

$$\left.\begin{aligned}
&\|x\,V_{i+1}(S, t, T) - x\,V_i(S, t, T)\| \\
&\qquad \le 2^{i+1}\,\psi_1(\tfrac{T-S}{2^{i+1}})\,\psi_2(\tfrac{T-S}{2^{i+1}})\exp\left(\tfrac{1}{2}\sum_{j=1}^{i}\mathcal{C}_j(T-S)\right)
\end{aligned}\right\} \qquad (11.7)$$

for

$$x \in B(6R). \qquad (11.8)$$

Proof. Let (11.8) be fulfilled. By (11.8) and (9.17)

$$x\,V_{i+1}(S,t,T) - x\,V_i(S,t,T) = 0 \quad \text{for } t \le \tfrac{S+T}{2}. \tag{11.9}$$

Let $\tfrac{S+T}{2} < t$. Then

$$x\,V_{i+1}(S,t,T) - x\,V_i(S,t,T) = \mathcal{U} + \mathcal{V} \tag{11.10}$$

where

$$\mathcal{U} = x\,V_i(S, \tfrac{S+T}{2}, \tfrac{S+T}{2})\,V_i(\tfrac{S+T}{2}, t, T)$$
$$- x\,V_{i-1}(S, \tfrac{S+T}{2}, \tfrac{S+T}{2})\,V_i(\tfrac{S+T}{2}, t, T),$$

$$\mathcal{V} = x\,V_{i-1}(S, \tfrac{S+T}{2}, \tfrac{S+T}{2})\,V_i(\tfrac{S+T}{2}, t, T)$$
$$- x\,V_{i-1}(S, \tfrac{S+T}{2}, \tfrac{S+T}{2})\,V_{i-1}(\tfrac{S+T}{2}, t, T)$$

the term $x\,V_i(S, \tfrac{S+T}{2}, \tfrac{S+T}{2})\,V_{i-1}(\tfrac{S+T}{2}, t, T)$ being well defined since $x\,V_i(S, \tfrac{S+T}{2}, \tfrac{S+T}{2}) \in \mathcal{B}(7R)$. By (11.8), Lemma 9.7 and Theorem 9.11

$$x \in \mathcal{B}(6R) \quad \text{and} \quad x\,V_{i-1}(S, \tfrac{S+T}{2}, \tfrac{S+T}{2}) \in \mathcal{B}(7R).$$

Hence by (11.5)

$$\|\mathcal{V}\| \le 2^i\,\psi_1(\tfrac{T-S}{2^{i+1}})\,\psi_2(\tfrac{T-S}{2^{i+1}}) \, \exp\Big(\tfrac{1}{2}\sum_{j=2}^{i} \mathcal{C}_j(T-S)\Big) \tag{11.11}$$

$$\left.\begin{aligned}
&\Big\| x\,V_i(S, \tfrac{S+T}{2}, \tfrac{S+T}{2}) - x\,V_{i-1}(S, \tfrac{S+T}{2}, \tfrac{S+T}{2}) \Big\| \\
&\qquad \le 2^i\,\psi_1(\tfrac{T-S}{2^{i+1}})\,\psi_2(\tfrac{T-S}{2^{i+1}}) \, \exp\Big(\tfrac{1}{2}\sum_{j=2}^{i} \mathcal{C}_j(T-S)\Big)
\end{aligned}\right\} \tag{11.12}$$

and by (11.12), (10.6), (9.18) and (9.8)

$$\|\mathcal{U}\| \le 2^i\psi_1(\tfrac{T-S}{2^{i+1}})\psi_2(\tfrac{T-S}{2^{i+1}})\Big(\exp\big(\tfrac{1}{2}\sum_{j=2}^{i}\mathcal{C}_j(T-S)\big)\Big)(1+\mathcal{C}_1(T-S)). \tag{11.13}$$

(11.10), (11.11) and (11.13) imply that

$$\|x\,V_{i+1}(S,t,T) - x\,V_i(S,t,T)\|$$
$$\le 2^i\psi_1(\tfrac{T-S}{2^{i+1}})\psi_2(\tfrac{T-S}{2^{i+1}})\Big(\exp\big(\tfrac{1}{2}\sum_{j=2}^{i}\mathcal{C}_j(T-S)\big)\Big)\,2\Big(1+\tfrac{1}{2}\mathcal{C}_1(\tfrac{T-S}{2})\Big)$$

and (11.7) is correct.   □

11.4. Theorem. *Let* $i \in \mathbb{N}$, $t \in [S,T] \subset [a,c]$ *and let* $x \in B(6R)$. *Then*

$$
\left.
\begin{aligned}
&\| x\, V_i(S,t,T) - x\, V_{i-1}(S,t,T) \| \\
&\quad \le 2^i\, \psi_1(\tfrac{T-S}{2^i})\, \psi_2(\tfrac{T-S}{2^i})\, \exp\Big(\tfrac{1}{2}\sum_{j=1}^{i-1} \mathcal{C}_j(T-S)\Big).
\end{aligned}
\right\} \tag{11.14}
$$

Proof. This is a consequence of Lemmas 11.2, 11.3.   □

11.5. Lemma. *Let* $x \in B(6R)$, $t \in [S,T] \subset [a,c]$, $k,\ell \in \mathbb{N}$, $k < \ell$. *Then*

$$
\left.
\begin{aligned}
&\| x\, V_\ell(S,t,T) - x\, V_k(S,t,T) \| \\
&\quad \le \sum_{i=k+1}^{\ell} 2^i\, \psi_1^2(\tfrac{c-a}{2^i})\, \exp\Big(\tfrac{1}{2}\sum_{j=1}^{\infty} \mathcal{C}_j(c-a)\Big).
\end{aligned}
\right\} \tag{11.15}
$$

Proof. The series

$$
\sum_{j=1}^{\infty} \mathcal{C}_j(c-a) = \sum_{j=1}^{\infty}\Big(\psi_1(\tfrac{c-a}{2^j}) + \mathcal{B}(\tfrac{c-a}{2^j}) \Big)
$$

is convergent by (7.1), (7.2), (7.4) and (7.21). Lemma 11.5 follows from Theorem 11.4.   □

11.6. Definition. Put

$$
x\, V(S,t,T) = \lim_{i \to \infty} x\, V_i(S,t,T) \tag{11.16}
$$

for $x \in B(6R)$, $t \in [S,T] \subset [a,c]$.

The limit in (11.16) exists by (11.15) since the series

$$
\sum_{i=1}^{\infty} 2^i\, \psi_1^2(\tfrac{c-a}{2^i})
$$

is convergent by (7.1), (7.2), (7.3).

11.7. Theorem. *Let* $x, \bar{x} \in B(6R)$, $t \in [S, T] \subset [a, c]$. *Then*

$$V \quad \text{is continuous}, \tag{11.17}$$

$$x\, V(S, t, T) \in B(7R), \tag{11.18}$$

$$x\, V(S, t, \tfrac{S+T}{2}) = x\, V(s, t, T) \quad \text{for} \ t \leq \tfrac{S+T}{2}, \tag{11.19}$$

$$\|x\, V(S, t, T) - x - x\, \mathbf{G}(S, t, T)\| \leq \mathcal{A}(2\,(T - S)), \tag{11.20}$$

$$\left.\begin{aligned}
\|x\, V(S, t, T) &- \bar{x}\, V(S, t, T) - x + \bar{x} \\
&- x\, \mathbf{G}(S, t, T) + \bar{x}\, \mathbf{G}(S, t, T)\| \\
&\leq \|x - \bar{x}\|\, \mathcal{B}(2(t - S)) .
\end{aligned}\right\} \tag{11.21}$$

Proof. V is the uniform limit of continuous V_i (cf. Lemma 11.5). Hence (11.17) is correct. (11.18) is a consequence of (9.23) and (9.1). (11.19) follows by (11.9). The limit procedure for $i \to \infty$ in (9.22) (cf. also (9.18)) gives

$$\left.\begin{aligned}
\|x\, V(S, t, T) - x - x\, \mathbf{G}(S, t, T)\| &\leq \mathcal{A}(T - S) \\
\text{for } x \in B(6R), \ t \in [S, T] &\subset [a, c] .
\end{aligned}\right\} \tag{11.22}$$

By (9.3)

$$x\, \mathbf{G}(S, t, \tfrac{S+T}{2}) = x\, \mathbf{G}(S, t, T)$$
$$\text{for} \ \ x \in B(8R), \ t \in [S, T] \subset [a, c], \ t \leq \tfrac{S+T}{2} .$$

Similarly

$$\left.\begin{aligned}
\|x\, V(S, t, T) - x - x\, \mathbf{G}(S, t, T)\| &\leq \mathcal{A}(\tfrac{T-S}{2^j}) \\
\text{for } x \in B(6R), \ t \leq S + \tfrac{T-S}{2^j}, \ j &\in \mathbb{N} .
\end{aligned}\right\} \tag{11.23}$$

(11.20) follows by (11.23) since $\mathcal{A}$ is nondecreasing and

$$2\,(t - S) \geq \tfrac{T-S}{2^j} \quad \text{for} \ \ \tfrac{T-S}{2^{j+1}} \leq t - S \leq \tfrac{T-S}{2^j} .$$

The limit procedure for $i \to \infty$ in (10.6) results in

$$\left.\begin{aligned}
\|x\, V(S, t, T) &- \bar{x}\, V(S, t, T) - x + \bar{x} \\
&- x\, \mathbf{G}(S, t, T) + \bar{x}\, \mathbf{G}(S, t, T)\| \\
&\leq \|x - \bar{x}\|\, \mathcal{B}(T - S) \\
\text{for } x \in B(6R), \ t \in [S, T] &\subset [a, c] .
\end{aligned}\right\} \tag{11.24}$$

(11.24) implies that

$$\left.\begin{array}{l} \|x\,V(S,t,T) - \bar{x}\,V(S,t,T) - x + \bar{x} \\ \qquad - x\,\mathbf{G}(S,t,T) + \bar{x}\,\mathbf{G}(S,t,T)\| \\ \qquad \leq \|x - \bar{x}\|\,\mathcal{B}(\tfrac{T-S}{2^j}) \\ \text{for } x, \bar{x} \in B(6R),\ t \leq S + \tfrac{T-S}{2^j},\ j \in \mathbb{N}, \end{array}\right\} \qquad (11.25)$$

in a similar way as (11.22) implies that (11.23) holds. (11.21) follows by (11.25) since $\mathcal{B}$ is nondecreasing. The proof is complete. □

11.8. Remark. Put $u(t) = x\,V(S,t,T)$, $u_i(t) = x\,V_i(S,t,T)$ for $t \in [S,T]$, $i \in \mathbb{N}$. (11.18) and (7.16) imply that u is an SR-solution of

$$\frac{\mathrm{d}}{\mathrm{d}t}x = G(x,\tau,t), \qquad (11.26)$$

$u(S) = x$. Therefore u_i may be called approximate solutions of (11.26).

Let $\mathrm{Dom}\,g \subset X \times \mathbb{R}$, $G(x,\tau,t) = g(x,\tau)\,t$. Then u_i are approximate solutions of the differential equation in the classical form

$$x = g(x,t)$$

which are obtained by Euler's method (going back to 1768).

Chapter 12

Solutions

12.1. Notation. By (7.1), (7.2), (7.4), (7.16), (7.18), (11.1) there exists $\xi_5 \in \mathbb{R}^+$ such that

$$\left. \begin{aligned} \xi_5 \le \xi_4, \quad \psi_1(\xi_5) + \mathcal{B}(2\,\xi_5) &\le \tfrac{1}{3}\,, \\ \psi_2(\xi_5) + \mathcal{A}(2\,\xi_5) &\le \tfrac{2}{3}\,R. \end{aligned} \right\} \tag{12.1}$$

Let $[a, c] \subset \mathbb{R}$,

$$c - a \le 2\,\xi_5, \quad b = \tfrac{1}{2}\,(a + c). \tag{12.2}$$

Let W be defined by

$$x\,W(S, t) = x\,V(S, t, S + b - a) \tag{12.3}$$

for $x \in B(6R)$, $S \in [a, b]$, $t \in [S, S + b - a]$. For $\sigma \in [S, S+b-a]$ put

$$\mathcal{E}_{S,\sigma} = \{x\,W(S, \sigma)\,; x \in B(5R)\}. \tag{12.4}$$

12.2. Lemma. W *is continuous.*

Proof. W is continuous since it is a restriction of a continuous V (cf. (11.17)). $\qquad\square$

12.3. Lemma. *Let* $S, t \in [a, b]$, $S \le t$, $v, \bar{v} \in B(6R)$. *Then*

$$\mathcal{E}_{S,t} \subset B(6R)\,, \tag{12.5}$$

$$\|v\,W(S, t) - v - v\,\mathbf{G}(S, t, S+b-a)\| \le \mathcal{A}(2\,(t - S))\,, \tag{12.6}$$

$$\left. \begin{aligned} &\|v\,W(S, t) - \bar{v}\,W(S, t) - v + \bar{v} - v\,\mathbf{G}(S, t, S+b-a) \\ &\quad + \bar{v}\,\mathbf{G}(S, t, S+b-a)\| \le \|v - \bar{v}\|\,\mathcal{B}(2(t - S))\,, \end{aligned} \right\} \tag{12.7}$$

$$\|v\,W(S, t) - v\| \le \psi_2(t - S) + \mathcal{A}(2(t - S))\,, \tag{12.8}$$

69

$$\|v\,W(S,t) - \bar{v}\,W(S,t) - v + \bar{v}\,\| \\ \leq \|v - \bar{v}\|\,(\psi_1(t - S) + \mathcal{B}(2\,(t - S)))\,, \qquad\qquad\Big\} \quad (12.9)$$

$$\|v\,W(S,t) - \bar{v}\,W(S,t)\| \\ \geq \|v - \bar{v}\|\,(1 - \psi_1(t - S) - \mathcal{B}(2\,(t{-}S)))\,. \qquad\quad\Big\} \quad (12.10)$$

Proof.　(12.5) is a consequence of (12.4) and (11.18). Further, (12.6) and (12.7) follow by (12.3), (11.20), (11.21). Moreover, (12.8) holds by (12.6) since

$$\|v\,W(S,t) - v\| \leq \|v\,\mathbf{G}(S,t,S + b - a)\| + \mathcal{A}(2(T - S)) \\ \leq \psi_2(T - S) + \mathcal{A}(2(T - S))$$

(cf. (9.7)). Similarly (12.9) holds by (12.7) and (9.8).

(12.10) follows from (12.9).　　　　　　　　　　　　　　　　　　　□

12.4. Lemma. *Let*

$$a \leq \tau \leq t \leq b. \qquad\qquad\qquad (12.11)$$

Then

$$x\,W(a,\tau)W(\tau,t) = x\,W(a,t), \quad x \in B(5R)\,. \qquad (12.12)$$

Proof.　$x\,W(a,\tau)\,W(\tau,t)$ is well defined since

$$x\,W(a,\tau) = x\,V(a,\tau,b) \in B(6R)$$

(cf. (12.3)). Let

$$i, j, k, \ell \in \mathbb{N}_0, \quad j \leq i, \quad k \leq \ell \leq 2^j\,, \\ \tau = a + k\,(b-a)\,2^{-j}, \quad t = a + \ell\,(b-a)\,2^{-j}\,. \qquad\Big\} \quad (12.13)$$

By Definition 9.4

$$x\,V_j(a,\tau,b)\,V_j(\tau,t,\tau + b - a) = x\,V_j(a,t,b)$$

and

$$x\,V_i(a,\tau,b)\,V_i(\tau,t,\tau + b - a) = x\,V_i(a,t,b) \\ \text{for } i \in \mathbb{N}, \; i > j\,. \qquad\qquad\Big\} \quad (12.14)$$

The limit procedure for $i \to \infty$ in (12.14) implies (cf. (12.3)) that (12.12) holds for the particular choice of τ, t in (12.13) since

$$\|v\,V_i(\tau, t, \tau + b - a) - \bar{v}\,V_i(\tau, t, \tau + b - a)\|$$

$$\leq \|v - \bar{v}\|\,(1 + \psi_1(T - S) + \mathcal{B}(T - S))$$

for v, $\bar{v} \in B(6R)$ (cf. (9.8), (10.6)).

Hence (12.12) holds since W is continuous and for every $i \in \mathbb{N}$, τ, $t \in [a, b]$, $\tau \leq t$ there exist k, $\ell \in \mathbb{N}_0$, $k \leq \ell$ such that

$$a + (k-1)(b-a)\,2^{-i} \leq \tau \leq a + k\,(b-a)\,2^{-i}$$

and

$$a + (\ell-1)\,(b-a)\,2^{-i} \leq t \leq a + \ell\,(b-a)\,2^{-i}. \qquad \square$$

12.5. Lemma. *Let $a \leq S \leq T \leq b$. Then*

$$W(S, T) \quad \text{is a one-to-one map of } B(5R) \text{ onto } \mathcal{E}_{S,T}. \tag{12.15}$$

Moreover, if $y \in B(5R)$ then

$$\text{there exists } x \in B(6R) \quad \text{such that } x\,W(a, S) = y. \tag{12.16}$$

Further

$$B(6R) \subset \mathcal{E}_{S,T}. \tag{12.17}$$

Proof. $W(S, T)$ is a map of $B(6R)$ onto $\mathcal{E}_{S,T}$ by (12.4) and (12.8). $W(S, T)$ is a one-to-one map by (12.10), (12.1).

Let $Q : B(6R) \to X$ be defined by

$$x\,Q = x\,W(S, T) - x.$$

Lemma B.1 may be applied with

$$R_1 = 6R, \quad R_2 = 5R, \quad \xi = \min\{\tfrac{1}{3}, \tfrac{2}{3}\,R\}.$$

Hence (cf. (B.3)) there exists $x \in B(6R)$ such that

$$x\,W(a, S) = x + x\,Q = y.$$

(12.16) holds. Finally, (12.17) is a consequence of (12.8). $\qquad \square$

12.6. Lemma. *Let $x \in B(6R)$, $u(t) = x\,W(a,t)$ for $t \in [a,b]$. Then*

$$\|u(\tau) - u(\bar{t})\| \le \psi_2(\tau - \bar{t}) + \mathcal{A}(2\,(\tau - \bar{t})) \tag{12.18}$$

and

$$\left.\begin{array}{l} \|u(t) - u(\bar{t}) - G(u(\tau), \tau, t) + G(u(\tau), \tau, \bar{t})\| \\[2mm] \quad \le \mathcal{A}(2(t - \bar{t}))\,(1 + \psi_1(t - \bar{t})) + 2\psi_1(t - \bar{t})\,\psi_2(t - \bar{t}) \end{array}\right\} \tag{12.19}$$

if

$$a \le \bar{t} \le \tau \le t \le b. \tag{12.20}$$

Proof.   Let (12.20) hold. Then (cf. (12.12))

$$u(t) = x\,W(a,t) = x\,W(a,\bar{t})\,W(\bar{t},t) = u(\bar{t})\,W(\bar{t},t).$$

By (12.6), (9.7)

$$\left.\begin{array}{l} \|u(t) - u(\bar{t}) - u(\bar{t})\,\mathbf{G}(\bar{t},t,\bar{t}+b-a)\| \le \mathcal{A}(2\,(t - \bar{t})), \\[2mm] \|u(\bar{t})\,\mathbf{G}(\bar{t},t,\bar{t}+b-a)\| \le \psi_2(t - \bar{t}) \end{array}\right\} \tag{12.21}$$

and (12.18) is correct. (9.3) and (9.2) imply that

$$u(\bar{t})\,\mathbf{G}(\bar{t},t,\bar{t}+b-a) = G(u(\bar{t}),\bar{t},t) - G(u(\bar{t}),\bar{t},\bar{t})$$

since

$$\widetilde{G}(x,\tau,t) = G(x,\tau,t) \quad \text{for } \tau,\, t \in [a,b],\ x \in B(8R).$$

(12.20) and (12.21) imply that

$$\|u(t) - u(\bar{t}) - G(u(\bar{t}),\bar{t},t) + G(u(\bar{t}),\bar{t},\bar{t})\| \le \mathcal{A}(2\,(t - \bar{t})). \tag{12.22}$$

Finally,

$$u(t) - u(\bar{t}) - G(u(\tau),\tau,t) + G(u(\tau),\tau,\bar{t}) = A + B + C \tag{12.23}$$

where

$$\begin{aligned} A &= u(t) - u(\bar{t}) - G(u(\bar{t}),\bar{t},t) + G(u(\bar{t}),\bar{t},\bar{t}), \\ B &= G(u(\bar{t}),\bar{t},t) - G(u(\bar{t}),\bar{t},\bar{t}) - G(u(\tau),\bar{t},t) + G(u(\tau),\bar{t},\bar{t}), \\ C &= G(u(\tau),\bar{t},t) - G(u(\tau),\bar{t},\bar{t}) - G(u(\tau),\tau,t) + G(u(\tau),\tau,\bar{t}) \end{aligned}$$

and (cf. (12.21), (8.4), (12.18), (8.6))

$$\|A\| \le \mathcal{A}(2(t-\bar{t})), \tag{12.24}$$

$$\begin{aligned}\|B\| &\le \|u(\tau) - u(\bar{t})\| \, \psi_1(\tau - \bar{t}) \\ &\le \psi_1(t-\bar{t})\,\psi_2(t-\bar{t}) + \mathcal{A}(2\,(t-\bar{t}))\,\psi_1(t-\bar{t}),\end{aligned}\Bigg\} \tag{12.25}$$

$$\|C\| \le \psi_2(t-\bar{t})\psi_1(t-\bar{t}). \tag{12.26}$$

(12.19) holds by (12.1), (12.23)–(12.26). □

12.7. Lemma. *Let* $x \in B(6R)$, $u(t) = xW(a,t)$ *for* $t \in [a,b]$, $\varepsilon > 0$. *Then there exists* $\xi > 0$ *such that*

$$\|u(t) - u(\bar{t}) - G(u(\tau),\tau,t) + G(u(\tau),\tau,\bar{t})\| \le \varepsilon\,\frac{t-\bar{t}}{b-a} \tag{12.27}$$

if

$$u(t) = x\,W(a,t) \quad for\ t \in [a,b], \tag{12.28}$$

$$a \le \bar{t} \le \tau \le t \le b,\ t - \bar{t} \le \xi. \tag{12.29}$$

Proof. (7.1)–(7.4) and (7.16), (7.19) imply that

$$\psi_1(\sigma) \to 0,\ \sigma^{-1}\mathcal{A}(\sigma) \to 0,\ \sigma^{-1}\psi_1(\sigma)\,\psi_2(\sigma) \to 0 \quad for\ \sigma \to 0+\ .$$

Hence Lemma 12.7 is a consequence of Lemma 12.6. □

12.8. Corollary. *u is an* SR-*solution of* (8.8).

12.9. Corollary.

$$\frac{\partial}{\partial t}(u(t) - G(u(\tau),\tau,t))|_{t=\tau} = 0 \quad for\ \tau \in [a,b].$$

Hence $\frac{d}{dt}u(t)|_{t=\tau}$ *exists if and only if* $\frac{\partial}{\partial t}G(u(\tau),\tau,t)|_{t=\tau}$ *exists.*

12.10. Theorem. *Let* $y \in B(4R)$, $s \in [a,b]$. *Then*

$$there\ exists\ x \in B(5R)\quad such\ that\ x\,W(a,s) = y. \tag{12.30}$$

Put $u(t) = x\,W(a,t)$ *for* $t \in [a,b]$. *Then*

$$u\quad is\ an\ \text{SR-}solution\ of\ (8.8)\ on\ [a,b], \tag{12.31}$$

$$\|u(t) - y\| \le R \quad for\ t \in [a,b]. \tag{12.32}$$

Moreover, x is unique.

Proof. (12.30) holds by (12.16), (12.8). Moreover, (12.31) is true by Corollary 12.8.

If $t < s$ then

$$y - u(t) = x\,W(a,t)\,W(t,s) - x\,W(a,t)$$

and $\|u(t) - y\| \le R$ by (12.8) and (12.1).

If $s < t$ then

$$u(t) - y = x\,W(a,s)\,W(s,t) - x\,W(a,s)$$

and again $\|u(t) - y\| \le R$ by (12.8). Hence (12.32) holds.

Let $\bar{x} \in B(5R)$ be such that $\bar{x}\,W(a,s) = y$. Then by (12.9), (12.1) and (12.2)

$$\|x - \bar{x}\| = \|x\,W(a,s) - \bar{x}\,W(a,s) - x + \bar{x}\|$$
$$\le \|x - \bar{x}\|\,(\psi_1(\xi_5) + \mathcal{B}(\xi_5)) \le \frac{1}{3}\,\|x - \bar{x}\|,$$

i.e. $x = \bar{x}$. $\qquad\square$

12.11. Theorem. *Let $[S,T] \subset [a,b]$ and let $w : [S,T] \to B(4R)$ be an SR-solution of (8.8). Then there exists a unique $z \in B(5R)$ such that*

$$w(s) = z\,W(a,s) \quad \text{for } s \in [S,T]. \tag{12.33}$$

Proof. Theorem 12.11 is a particular case of Theorem 16.1 since every SKH-solution of (8.8) is an SR-solution of (8.8). $\qquad\square$

12.12. Theorem. *Let $y \in B(3R)$, $s \in [S,T] \subset [a,b]$.*

Let $v : [S,T] \to B(8R)$ be an SR-solution of (8.8), $v(s) = y$.

Then

$$\|v(t)\| \le \|y\| + \tfrac{2}{3} R \quad \text{for } t \in [S,T] \tag{12.34}$$

and

$$\left.\begin{array}{l} \text{there exists a unique } z \in B(\|y\| + \tfrac{2}{3} R) \\[4pt] \text{such that } v(t) = z\,W(a,t) \ \text{ for } t \in [S,T]. \end{array}\right\} \tag{12.35}$$

Proof. Put

$$\widehat{S} = \inf\{S_1\,;\, S \le S_1 \le s,\, \|v(t)\| \le 4R \text{ for } S_1 \le t \le s\},$$
$$\widehat{T} = \sup\{T_1\,;\, s \le T_1 \le T,\, \|v(t)\| \le 4R \text{ for } s \le t \le T_1\}.$$

Then $\widehat{S} < s < \widehat{T}$ since v is continuous by Lemma 6.5. By Theorem 12.11 there exists $z \in B(5R)$ such that $v(t) = z\,W(a,t)$ for $t \in [\widehat{S}, \widehat{T}]$.

If $\widehat{S} \leq t \leq s$ write

$$v(t) - y = z\,W(a,t) - z\,W(a,t)\,W(t,s)$$

and (cf. (12.8), (12.1))

$$\|v(t) - y\| \leq \psi_2(s-t) + \mathcal{A}(2s - 2t) \leq \frac{2}{3}\,R.$$

Hence

$$\|v(t)\| \leq \|y\| + \frac{2}{3}\,R \leq \frac{11}{3}\,R,$$

which implies that

$$\widehat{S} = S \quad \text{and} \quad \|v(t)\| \leq \|y\| + \frac{2}{3}\,R \quad \text{for} \quad S \leq t \leq s. \tag{12.36}$$

If $s \leq t \leq T$ write

$$v(t) - y = z\,W(a,s)\,W(s,t) - z\,W(a,s)$$

and by a similar argumentation

$$\widehat{T} = T \quad \text{and} \quad \|v(t)\| \leq \|y\| + \frac{2}{3}\,R \quad \text{for} \quad s \leq t \leq T. \tag{12.37}$$

Now, (12.34) and (12.35) hold by (12.36) and (12.37). The proof is complete. $\qquad\square$

Chapter 13

Continuous dependence

13.1 . Notation. Let (12.1), (12.2) hold and let $\eta > 0$. Assume that $G : B(8R) \times [a,b]^2 \to X$ and $G^* : B(8R) \times [a,b]^2 \to X$ fulfil (8.2)–(8.7),

$$\|G^*(x, \tau, t) - G(x, \tau, t)\| \leq \eta \quad \text{for } x \in B(8R), \ \tau, \ t \in [a,b]. \qquad (13.1)$$

Starting from G^* introduce $\widetilde{G}^*$, $\mathbf{G}^*$, V_i^*, V^*, W^*, $\mathcal{E}_{S,T}^*$ in the same way as $\widetilde{G}$, $\mathbf{G}$, V_i, V, W, $\mathcal{E}_\sigma$ were introduced starting from G.

13.2. Lemma. *Let* $[S, S + \tau] \subset [a,b]$, $z \in B(6R)$. *Then*

$$\|z\, W(S, S+\tau) - z\, W^*(S, S+\tau)\| \leq \eta + 2\, \mathcal{A}(2\tau). \qquad (13.2)$$

Proof. (13.2) holds by (12.6), (9.3) and by an analogous inequality for W^*. $\qquad\qquad\qquad\qquad\qquad\qquad\qquad\qquad\qquad\qquad \Box$

13.3. Lemma. *Assume that* $\ell \in \mathbb{N}_0$ *and that*

$$
\left.
\begin{aligned}
&\|z\, W(\widehat{S}, \widehat{S} + 2^\ell \tau) - z\, W^*(\widehat{S}, \widehat{S} + 2^\ell \tau)\| \\
&\leq (\eta + 2\, \mathcal{A}(2\tau)) 2^{\ell-1} \exp\left(\sum_{j=1}^{\ell-1} (\psi_1(2^j\, \tau) + \mathcal{B}\,(2^{j+1}\, \tau)) \right)
\end{aligned}
\right\} \qquad (13.3)
$$

for $[\widehat{S}, \widehat{S} + 2^\ell\, \tau] \subset [a,b]$ *and* $z \in B(5R)$, *where for the case* $\ell = 1$ *we take*

$$\sum_{j=1}^{0} (\psi_1(2^j\, \tau) + \mathcal{B}\,(2^{j+1}\, \tau)) = 0.$$

77

Then

$$\|x\, W(S, S + 2^{\ell+1}\,\tau) - x\, W^*(S, S + 2^{\ell+1}\,\tau)\|$$
$$\le (\eta + 2\,\mathcal{A}(2\tau))\, 2^{\ell} \exp\left(\sum_{j=1}^{\ell} (\psi_1(2^j\tau) + \mathcal{B}\,(2^{j+1}\tau))\right) \tag{13.4}$$

for $[S, S + 2^{\ell+1}\tau] \subset [a, b]$, $x \in B(5R)$.

Proof. Let

$$[S, S + 2^{\ell+1}\tau] \subset [a, b], \quad x \in B(5R)\,.$$

Then (cf. (12.12))

$$x\, W(S, S + 2^{\ell+1}\tau) - x\, W^*(S, S + 2^{\ell+1}\tau) = J + K\,, \tag{13.5}$$

where

$$J = x\, W(S, S + 2^{\ell}\tau)\, W(S + 2^{\ell}\tau, S + 2^{\ell+1}\tau)$$
$$- x\, W(S, S + 2^{\ell}\tau)\, W^*(S + 2^{\ell}, S + 2^{\ell+1}\tau)\,,$$

$$K = x\, W(S, S + 2^{\ell}\tau)\, W^*(S + 2^{\ell}\tau, S + 2^{\ell+1}\tau)$$
$$- x\, W^*(S, S + 2^{\ell}\tau)\, W^*(S + 2^{\ell}\tau, S + 2^{\ell+1}\tau)\,.$$

Observe that $x\, W(S, S + 2^{\ell}\tau)\, W^*(S + 2^{\ell}\tau, S + 2^{\ell+1}\tau)$ is well defined since $x\, W(S, S + 2^{\ell}\tau) \subset B(6R)$. By (13.3)

$$\|J\| \le (\eta + 2\,\mathcal{A}(2\tau))\, 2^{\ell-1} \exp\left(\sum_{j=1}^{\ell-1} (\psi_1(2^j\tau) + \mathcal{B}\,(2^{j+1}\,\tau))\right) \tag{13.6}$$

and by (13.3), (12.9)

$$\|K\| \le (\eta + 2\,\mathcal{A}(2\,\tau))\, 2^{\ell-1} \left[\exp\left(\sum_{j=1}^{\ell-1} (\psi_1(2^j\tau) + \mathcal{B}\,(2^{j+1}\tau))\right)\right] \times$$
$$\times \left(1 + \psi_1(2^{\ell}\tau) + \mathcal{B}\,(2^{\ell+1}\tau)\right)\,. \tag{13.7}$$

(13.4) is a consequence of (13.5)–(13.7). □

13.4. Lemma. *Let* $[S, T] \subset [a, b]$, $x \in B(5R)$, $\ell \in \mathbb{N}$. *Then*

$$\|x\, W(S, T) - x\, W^*(S, T)\|$$
$$\le (\eta + 2\,\mathcal{A}(2^{-\ell+1}\,(T - S)))\, 2^{\ell}\, \varkappa\,, \tag{13.8}$$

where

$$\varkappa = \exp \sum_{j=1}^{\infty} (\psi_1((b-a)\,2^{-j}) + \mathcal{B}\,((b-a)\,2^{-j+1}))\,. \tag{13.9}$$

Proof. Lemmas 13.2 and 13.3 imply that

$$\left.\begin{aligned}&\|x\,W(S,S+2^{\ell}\tau) - x\,W^*(S,S+2^{\ell}\tau)\|\\&\qquad\leq (\eta + 2\,\mathcal{A}(2\,\tau))\,2^{\ell-1} \exp \sum_{j=1}^{\ell-1} (\psi_1(2^j\tau) + \mathcal{B}\,(2^{j+1}\tau))\end{aligned}\right\} \tag{13.10}$$

for $\ell \in \mathbb{N}$, $[S, S + 2^{\ell}\tau] \subset [a,b]$, $x \in \mathcal{E}_S$.

(13.8) is a consequence of (13.10) since the series in (13.9) is convergent by (7.2), (7.4), (7.18) and $\tau \leq (b-a)\,2^{-\ell}$. $\qquad\square$

13.5. Lemma. *Define*

$$\Omega(\eta) = \min \left\{ [\eta + 2\,\mathcal{A}(2\,(b-a)2^{-\ell})]\,2^{\ell}; \ell \in \mathbb{N}_0 \right\} \quad for\ \eta > 0\,. \tag{13.11}$$

Then

$$\Omega(\eta) \to 0 \quad for\ \eta \to 0+\,. \tag{13.12}$$

Proof. Let $\varepsilon > 0$. By (7.16)

$$2^{\ell+1}\,\mathcal{A}(2(b-a)2^{-\ell}) \to 0 \quad for\ \ell \to \infty$$

and there exists $m \in \mathbb{N}$ such that

$$2^{m+1}\,\mathcal{A}(2\,(b-a)\,2^{-m}) < \tfrac{1}{2}\,\varepsilon\,. \tag{13.13}$$

Let

$$0 < \eta \leq 2^{-m-1}\,\varepsilon\,.$$

Then

$$[\eta + 2\,\mathcal{A}(2\,(b-a)\,2^{-m})]\,2^m < \varepsilon\,.$$

Hence (13.12) is valid. $\qquad\square$

13.6. Lemma. *Let* $S \in [a,b]$, $y \in B(5R)$. *Then*

$$\|y\,W(S,T) - y\,W^*(S,T)\| \leq \Omega(\eta)\,\varkappa \quad for\ T \in [S,b]\,. \tag{13.14}$$

Proof. (13.14) is a consequence of (13.8), (13.11), (13.12) since $\mathcal{A}$ is nondecreasing. $\qquad\square$

13.7 . Theorem. *Let* $a \leq S \leq s \leq T \leq b$, $S < T$, $y \in B(3R)$. *Assume that* $v : [S, T] \to B(8R)$ *is an SR-solution of*

$$\frac{\mathrm{d}}{\mathrm{d}t} x = \mathrm{D}_t G(x, \tau, t), \qquad (13.15)$$

$v(s) = y$ *and that* $v^* : [S, T] \to B(8R)$ *is an SR-solution of*

$$\frac{\mathrm{d}}{\mathrm{d}t} x = \mathrm{D}_t G^*(x, \tau, t), \qquad (13.16)$$

$v^*(s) = y$. *Then*

$$\|v(t) - v^*(t)\| \leq \tfrac{3}{2} \Omega(\eta) \varkappa \quad \textit{for } t \in [S, T]. \qquad (13.17)$$

Proof. By Theorem 12.12 there exist z, $z^* \in B(\|y\| + \tfrac{2}{3}R)$ such that

$$v(t) = z\,W(a, t), \quad v^*(t) = z^*\,W^*(a, t) \quad \text{for } t \in [S, T]. \qquad (13.18)$$

Lemma 13.6 implies that

$$\left. \begin{aligned} \|q\,W(\sigma, \tau) - q\,W^*(\sigma, \tau)\| &\leq \Omega(\eta)\varkappa \\ \text{for } q \in B(5R),\ a &\leq \sigma \leq \tau \leq b. \end{aligned} \right\} \qquad (13.19)$$

If $s \leq t \leq T$ then (cf. Lemma 13.6)

$$\|v(t) - v^*(t)\| = \|y\,W(s, t) - y\,W^*(s, t)\| \leq \Omega(\eta)\varkappa. \qquad (13.20)$$

Let $S \leq t < s$. Then (cf. (13.18), (12.12))

$$y = z\,W(a, s) = z\,W(a, t)\,W(t, s) = v(t)\,W(t, s)$$

and similarly

$$y = v^*(t)\,W^*(t, s).$$

Hence

$$\left. \begin{aligned} 0 &= -v(t)\,W(t, s) + v^*(t)W(t, s) \\ &\quad -v^*(t)\,W(t, s) + v^*(t)\,W^*(t, s), \\ \|v(t) - v^*(t)\| &\leq \|v(t) - v^*(t) - v(t)W(t, s) + v^*(t)W(t, s)\| \\ &\quad + \|v^*(t)\,W(t, s) - v^*(t)\,W^*(t, s)\|. \end{aligned} \right\} \qquad (13.21)$$

By (12.9) and (12.1)

$$\|v(t) - v^*(t) - v(t)\,W(t, s) + v^*(t)\,W(t, s)\| \leq \|v(t) - v^*(t)\| \tfrac{1}{3}, \qquad (13.22)$$

by (13.14)

$$\|v^*(t)\,W(t,s) - v^*(t)\,W^*(t,s)\| \le \Omega(\eta)\,\varkappa. \tag{13.23}$$

(13.21)–(13.23) imply that

$$\|v(t) - v^*(t)\| \le \tfrac{3}{2}\,\Omega(\eta)\,\varkappa. \tag{13.24}$$

(13.17) holds by (13.20) and (13.24) and Theorem 13.7 is true.

$\square$

13.8. Remark. The problem of the necessity of the condition

$$\sum_{i=1}^{\infty} 2^i\,\psi_1(2^{-i})\,\psi_2(2^{-i}) < \infty \tag{13.25}$$

in the above theory was studied by J. Jarník (cf. [Jarník (1961a)],[Jarník (1961b)]).

Assume that for ψ_1, ψ_2 we have

$$\psi_1 = \psi_2, \tag{13.26}$$

while ψ_1, ψ_2 fulfil (7.1), (7.2), (7.4),

$$\psi_1(\sigma_1 + \sigma_2) \le \psi_1(\sigma_1) + \psi_1(\sigma_2) \quad \text{for } \sigma_1, \sigma_2 \ge 0 \tag{13.27}$$

and

$$\sum_{i=1}^{\infty} 2^i\,\psi_1(2^{-i})\,\psi_2(2^{-i}) = \infty. \tag{13.28}$$

Then there exists a sequence of continuous functions $a_k : [0,1] \to \mathbb{R}$, $b_k : [0,1] \to \mathbb{R}$, $k \in \mathbb{N}$, such that the functions

$$G_k(x,t) = x \int_0^t a_k(s)\,\mathrm{d}s + \int_0^t b_k(s)\,\mathrm{d}s \quad t \in [0,1],\ |x| \le 1$$

fulfil

$$|G_k(x,t_2) - G_k(x,t_1)| \le \psi_1(|t_2 - t_1|),$$

$$|\Delta_v\,(G_k(x,t_2) - G_k(x,t_1))| \le \|v\|\,\psi_1(|t_2 - t_1|)$$

for x, $x + v \in [-1,1]$, $t_1, t_2 \in [0,1]$, $G_k(x,t) \to 0$ uniformly for $k \to \infty$. Let u_k be a solution of

$$\frac{\mathrm{d}}{\mathrm{d}t} x = D_t G_k(x,t), \quad u_k(0) = 0.$$

The sequence u_k is divergent. If conditions (13.26), (13.27) are added to the theory which is presented in Chapters 8–13 then condition (13.25) becomes necessary. Observe that u_k is also a solution of

$$\dot{x} = x\, a_k(t) + b_k(t).$$

For details see [Jarník (1961a)].

On the other hand, there exist couples ψ_1, ψ_2 such that (7.1), (7.2), (7.4), (13.28) are fulfilled and the continuous dependence holds. This occurs if

$$\frac{\psi_1(\sigma)}{\sigma} \to 0 \quad \text{for} \quad \sigma \to 0+$$

very slowly, e.g. if

$$\frac{\psi_1(\sigma)}{\sigma\,|\ln\sigma|} \to 1 \quad \text{for} \quad \sigma \to 0+\,.$$

See [Jarník (1961b)].

Chapter 14

Strong Kurzweil Henstock-integration of functions of a pair of coupled variables

14.1. Notation. $\operatorname{Dom} U \subset \mathbb{R}^2$, $U : \operatorname{Dom} U \to X$, $[a, b] \subset \mathbb{R}$, $\delta_0 : [a, b] \to \mathbb{R}^+$, $\{(\tau, t) ; \tau - \delta_0(\tau) \le t \le \tau + \delta_0(\tau)\} \subset \operatorname{Dom} U$, $u : [a, b] \to X$.

If $a \le \sigma < b$ and if $\lim\limits_{\tau \to \sigma, \tau > \sigma} u(\tau)$ exists, denote it by $u(\sigma+)$. Similarly, $u(\sigma-) = \lim\limits_{\tau \to \sigma, \tau < \sigma} u(\tau)$ if $a < \sigma \le b$ and if the limit exists.

14.2. Definition. A couple $([\bar{t}, t], \tau)$ or $([\bar{s}, s], \sigma)$ is called a *tagged interval*, τ, σ being the *tags*. A finite set

$$\mathcal{A} = \{([\bar{s}_i, s_i], \sigma_i) ; i = 1, 2, \ldots, k\} \tag{14.1}$$

is called a *system in* $[a, b]$ if

$$\sigma_i \in [\bar{s}_i, s_i] \subset [a, b] \quad \text{for} \quad i = 1, 2, \ldots, k$$

and if the intervals $[\bar{s}_i, s_i]$, $[\bar{s}_j, s_j]$ are nonoverlapping for $i \ne j$ (i.e. their intersections $[\bar{s}_i, s_i] \cap [\bar{s}_j, s_j]$ contain at most one point).

Let $\delta : [a, b] \to \mathbb{R}^+$. The system A is δ-*fine* if

$$[\bar{t}, t] \subset [\tau - \delta(\tau), \tau + \delta(\tau)] \quad \text{for} \quad ([\bar{t}, t], \tau) \in A.$$

If $\xi \in \mathbb{R}^+$ and if $t - \bar{t} \le \xi$ for $([\bar{t}, t], \tau) \in \mathcal{A}$, then A is called ξ-*fine*.

Let $Q \subset [a, b]$. The system $\mathcal{A}$ is called Q-*anchored* if $\tau \in Q$ for $([\bar{t}, t], \tau) \in \mathcal{A}$.

The system $\mathcal{A}$ is called a *partition* of $[a, b]$ if

$$\bigcup_{i=1}^{k} [\bar{s}_i, s_i] = [a, b]. \tag{14.2}$$

14.3. Remark. Any partition $\mathcal{A}$ of $[a, b]$ can be written in the form

$$\mathcal{A} = \{([t_{i-1}, t_i], \tau_i) ; i = 1, 2, \ldots, k\} \tag{14.3}$$

where

$$a = t_0 \leq \tau_1 \leq t_1 \leq \tau_2 \leq t_2 \leq \cdots \leq t_{k-1} \leq \tau_k \leq t_k = b. \tag{14.4}$$

The index i is frequently omitted so that the system $\mathcal{A}$ is written in the form

$$\mathcal{A} = \{([\bar{t}, t], \tau)\} \tag{14.5}$$

and e.g. (14.2) assumes the form

$$\bigcup_{\mathcal{A}} [\bar{t}, t] = [a, b]. \tag{14.6}$$

14.4. Lemma. (Cousin) *Given* $\delta : [a, b] \to \mathbb{R}^+$, *there exists a δ-fine partition* $\mathcal{A} = \{([\bar{t}, t], \tau)\}$ *of* $[a, b]$.

Lemma 14.4 can be found in almost all books on Kurzweil–Henstock integration. The proof can be based on the Heine–Borel covering theorem (e.g. [Kurzweil (1957)], Lemma 1.1.1; [Henstock (1988)], Theorem 3.1) or on the repeated bisection (e.g. [Henstock (1988)], Theorem 4.1; [Pfeffer (1993)], Proposition 1.2.4; [Kurzweil (2000)], Lemma 13.4) or on the concept of supremum (e.g. [Bartle, Sherbert (2000)], Theorem 5.5.5).

The SKH-integral (Strong Kurzweil–Henstock) which is introduced in Definitions 14.5 and 14.10, and the SR-integral (Strong Riemann) are useful in the theory of GODEs.

14.5. Definition. U is SKH-*integrable* (Strongly Kurzweil–Henstock integrable) on $[S, T]$ and u is an SKH-*primitive* of U on $[S, T]$ if

$$\left.\begin{array}{l} \text{for every } \varepsilon > 0 \text{ there exists } \delta : [S, T] \to \mathbb{R}^+ \text{ such that} \\ \displaystyle\sum_{i=1}^{k} \|u(t_i) - u(t_{i-1}) - U(\tau_i, t_i) + U(\tau_i, t_{i-1})\| \leq \varepsilon \end{array}\right\} \tag{14.7}$$

for every δ-fine partition $\mathcal{A} = \{([t_{i-1}, t_i], \tau_i); i = 1, 2, \ldots, k\}$ of $[S, T]$. Briefly, U is called SKH-*integrable* and u is called a *primitive* of U. The couple (τ, t) is called a *pair of coupled variables* .

14.6. Remark. If $\mathcal{A}$ is written in the form (14.5) then the inequality in (14.7) is replaced by

$$\sum_{\mathcal{A}} \|u(t) - u(\bar{t}) - U(\tau, t) + U(\tau, \bar{t})\| \leq \varepsilon.$$

14.7. Lemma. *Assume that U is SKH-integrable on $[a, b]$, u is its primitive and $\varepsilon > 0$. Let $\delta : [a, b] \to \mathbb{R}^+$ correspond to ε by Definition 14.5 and let $[S, T] \subset [a, b]$. Then*

$$\sum_{\mathcal{A}} \|u(t) - u(\bar{t}) - U(\tau, t) + U(\tau, \bar{t})\| \leq \varepsilon \tag{14.8}$$

for every δ-fine partition $\mathcal{A} = \{([\bar{t}, t], \tau)\}$ of $[S, T]$.

Proof. The proof is analogous to the proof of Lemma 5.3.

Let $\mathcal{A}$ be a δ-fine partition of $[S, T]$. By Lemma 14.4 there exist δ-fine partitions $\mathcal{B} = \{([\bar{t}, t], \tau)\}$ of $[a, S]$ and $\mathcal{C} = \{([\bar{t}, t], \tau)\}$ of $[T, b]$. Then

$$\sum_{\mathcal{A}} \|u(t) - u(\bar{t}) - U(\tau, t) + U(\tau, \bar{t})\|$$

$$\leq \sum_{\mathcal{A} \cup \mathcal{B} \cup \mathcal{C}} \|u(t) - u(\bar{t}) - U(\tau, t) + U(\tau, \bar{t})\| \leq \varepsilon$$

since $\mathcal{A} \cup \mathcal{B} \cup \mathcal{C}$ is a δ-fine partition of $[a, b]$. (14.8) is correct.

$\square$

14.8. Corollary. *Let U be SKH-integrable on $[a, b]$, u being its primitive, $[S, T] \subset [a, b]$. Then U is SKH-integrable on $[S, T]$ and u is its primitive.*

14.9. Lemma. *Let U be SKH-integrable on $[a, b]$, u being its primitive. Then*

(i) *if $y \in X$, $v(t) = u(t) + y$ for $t \in [a, b]$, then v is a primitive of U;*

(ii) *if v is a primitive of U, then $v(t) - u(t) = v(a) - u(a)$ for $t \in [a, b]$.*

The proof runs along the same lines as the proof of Lemma 5.5.

14.10. Definition. Let U be SKH-integrable on $[S, T]$ and let u be its primitive. Put

$$(\text{SKH}) \int_S^T D_t U(\tau, t) = u(T) - u(S). \tag{14.9}$$

(SKH) $\int_S^T D_t U(\tau, t)$ is called the *strong Kurzweil–Henstock integral* of U over $[S, T]$ or shortly, the SKH-*integral* of U over $[S, T]$.

14.11. Remark. By Lemma 14.9, (SKH) $\int_S^T D_t U(\tau, t)$ is independent of the choice of the primitive of U.

14.12. Remark. If $\xi > 0$ and $\delta : [S, T] \to \mathbb{R}^+$, $\delta(\tau) \leq \xi/2$ for $\tau \in [S, T]$ then every δ-fine partition $\mathcal{A} = \{([\bar{t}, t], \tau)\}$ of $[S, T]$ is a ξ-fine partition of $[S, T]$. Hence every SR-integrable U is SKH-integrable and

$$(\text{SKH}) \int_S^T D_t U(\tau, t) = (\text{SR}) \int_S^T D_t U(t, \tau).$$

14.13. Lemma. *Let $a < c < b$ and let the integrals*

$$(\text{SKH}) \int_a^c D_t U(\tau, t), \quad (\text{SKH}) \int_c^b D_t U(\tau, t) \qquad (14.10)$$

exist.

Then U is SKH-integrable on $[a, b]$ and

$$(\text{SKH}) \int_a^b D_t U(\tau, t) = (\text{SKH}) \int_a^c D_t U(\tau, t) + (\text{SKH}) \int_c^b D_t U(\tau, t). \qquad (14.11)$$

Proof. Let $w_1 : [a, c] \to X$ be a primitive of U on $[a, c]$, let $w_2 : [c, b] \to X$ be a primitive of U on $[c, b]$ and let $\varepsilon > 0$. By Lemma 14.4 and Definition 14.5 there exists $\delta : [a, b] \to \mathbb{R}^+$ such that

$$\tau + \delta(\tau) < c \text{ for } \tau < c, \quad \tau - \delta(\tau) > c \text{ for } \tau > c, \qquad (14.12)$$

$$\sum_{\mathcal{A}} \| w_1(t) - w_1(\bar{t}) - U(\tau, t) + U(\tau, \bar{t}) \| \leq \varepsilon \qquad (14.13)$$

for every δ-fine partition $\mathcal{A} = \{([\bar{t}, t], \tau)\}$ of $[a, c]$ and

$$\sum_{\mathcal{B}} \| w_2(t) - w_2(\bar{t}) - U(\tau, t) + U(\tau, \bar{t}) \| \leq \varepsilon \qquad (14.14)$$

for every δ-fine partition $\mathcal{B} = \{([\bar{t}, t], \tau)\}$ of $[c, b]$. Put

$$u(t) = \begin{cases} w_1(t) & \text{for } a \leq t \leq c, \\ w_2(t) - w_2(c) + w_1(c) & \text{for } c < t \leq b. \end{cases}$$

Let $\mathcal{C} = \{([\bar{t}, t], \tau)\}$ be a δ-fine partition of $[a, b]$. (14.12) implies that either

$$\left.\begin{array}{c} \text{there exist } \bar{t} < c \text{ and } t > c \text{ such that} \\ ([\bar{t}, c], c) \in \mathcal{C}, \ \ ([c, t], c) \in C \end{array}\right\} \qquad (14.15)$$

or

$$\left.\begin{array}{c} \text{there exist } \bar{s} < c \text{ and } s > c \text{ such that} \\ ([\bar{s}, s], c) \in \mathcal{C}. \end{array}\right\} \qquad (14.16)$$

If (14.15) holds then

$$\left.\begin{aligned} &\sum_{\mathcal{C}} \|u(t) - u(\bar{t}) - U(\tau, t) + U(\tau, \bar{t})\| \\ &= \sum_{\mathcal{C}; [\bar{t}, t] \subset [a, c]} \|w_1(t) - w_1(\bar{t}) - U(\tau, t) + U(\tau, \bar{t})\| \\ &\quad + \sum_{\mathcal{C}; [\bar{t}, t] \subset [c, b]} \|w_2(t) - w_2(\bar{t}) - U(\tau, t) + U(\tau, \bar{t})\| \\ &\leq \varepsilon + \varepsilon. \end{aligned}\right\} \qquad (14.17)$$

If (14.16) takes place then

$$\begin{aligned} u(s) - u(\bar{s}) &- U(c, s) + U(c, \bar{s}) \\ &= w_2(s) - w_2(c) + w_1(c) - w_1(\bar{s}) \\ &\quad - U(c, s) - U(c, c) + U(c, c) + U(c, \bar{s}), \end{aligned}$$

$$\begin{aligned} \|u(s) - u(\bar{s}) &- U(c, s) + U(c, \bar{s})\| \\ &\leq \|w_2(s) - w_2(c) - U(c, s) + U(c, c)\| \\ &\quad + \|w_1(c) - w_1(\bar{s})\| - U(c, c) + U(c, \bar{s})\|. \end{aligned}$$

Therefore

$$\left.\begin{aligned} &\sum_{\mathcal{C}} \|u(t) - u(\bar{t}) - U(\tau, t) + U(\tau, \bar{t})\| \\ &\leq \Big[\sum_{\mathcal{C}; [\bar{t}, t] \subset [a, c]} \|w_1(t) - w_1(\bar{t}) - U(\tau, t) + U(\tau, \bar{t})\| \\ &\quad + \|w_1(c) - w_1(\bar{s}) - U(c, c, + U(c, \bar{s})\| \Big] \\ &\quad + \Big[\|w_2(s) - w_2(c) - U(c, s) + U(c, c)\| \\ &\quad + \sum_{\mathcal{C}; [\bar{t}, t] \subset [c, b]} \|w_2(t) - w_2(\bar{t}) - U(\tau, t) + U(\tau, \bar{t})\| \Big] \\ &\leq 2\varepsilon. \end{aligned}\right\} \qquad (14.18)$$

By (14.17) and (14.18)

$$\sum_{\mathcal{C}} \|u(t) - u(\bar{t}) - U(\tau,t) + U(\tau,\bar{t})\| \leq 2\,\varepsilon.$$

U is SKH-integrable on $[a,b]$ and u is its primitive.   □

14.14. Remark. Let H, $\widehat{H}:[-1,1] \to \{0,1\}$ be defined by

$$H(t) = \begin{cases} 0 & \text{for } -1 \leq t \leq 0, \\ 1 & \text{for } \quad 0 < t \leq 1, \end{cases} \qquad \widehat{H}(t) = \begin{cases} 0 & \text{for } -1 \leq t < 0, \\ 1 & \text{for } \quad 0 \leq t \leq 1. \end{cases}$$

Then

$$(\mathrm{R}) \int_{-1}^{0} \widehat{H} \, dH = 0, \quad (\mathrm{R}) \int_{0}^{1} \widehat{H} \, dH = 1,$$

but $(\mathrm{R}) \displaystyle\int_{-1}^{1} \widehat{H} \, dH$ does not exist. On the other hand,

$$(\mathrm{SKH}) \int_{-1}^{1} D_t(\widehat{H}(\tau) \, H(t)) = 1$$

(cf. Lemma 14.13 and Remark 14.12). Of course,

$$(\mathrm{SR}) \int_{-1}^{0} D_t(\widehat{H}(\tau) \, H(t)) = (\mathrm{R}) \int_{-1}^{0} \widehat{H}(t) \, dH(t) = (\mathrm{R}) \int_{-1}^{0} \widehat{H} \, dH \quad \text{etc.}$$

14.15. Theorem. *The three conditions below are equivalent:*

(i) $u(T) - u(S) = (\mathrm{SKH}) \displaystyle\int_{S}^{T} D_t U(\tau,t)$ *for* $[S,T] \subset [a,b]$,

(ii) U *is SKH-integrable on* $[a,b]$ *and* u *is its primitive*,

(iii) *for* $\varepsilon > 0$ *there exists* $\delta:[a,b] \to \mathbb{R}^+$ *such that*

$$\sum_{\mathcal{A}} \|u(t) - u(\bar{t}) - U(\tau,t) + U(\tau,\bar{t})\| \leq \varepsilon$$

if $\mathcal{A} = \{([\bar{t},t],\tau)\}$ *is a* δ-*fine partition of* $[a,b]$.

Proof. Let (i) hold. Then $(\mathrm{SKH}) \int_S^T D_t U(\tau,t)$ exists for $[S,T] \subset [a,b]$, i.e. U is SKH-integrable on $[a,b]$ and $u:[a,b] \to X$ is an SKH-primitive of U. Hence (ii) is valid.

Let (ii) hold. Then (iii) is valid by Definition 14.5.

Let (iii) hold. Then (i) is true by Definitions 14.5 and 14.10.   □

14.16. Remark. Let $a \leq S < T \leq b$ and let U be SKH-integrable on $[S, T]$, u being its primitive. It is convenient to put

$$\left. \begin{array}{l} (\text{SKH}) \displaystyle\int_T^S D_t U(\tau, t) = -(\text{SKH}) \displaystyle\int_S^T D_t U(\tau, t) = u(S) - u(T), \\[3mm] (\text{SKH}) \displaystyle\int_S^S D_t U(\tau, t) = 0. \end{array} \right\} \tag{14.19}$$

14.17. Lemma. *Let U be SKH-integrable on $[a, b]$, $u : [a, b] \to X$ being its primitive. Then*

$$u(T) - u(\sigma) - U(\sigma, T) + U(\sigma, \sigma) \to 0 \quad \text{for} \quad T \to \sigma. \tag{14.20}$$

Proof. Let $\varepsilon > 0$ and let $\delta : [a, b] \to \mathbb{R}^+$ correspond to ε by Theorem 14.15 (iii). Let $a \leq \sigma < T \leq \sigma + \delta(\sigma) \leq b$. By Lemma 14.4 there exists a δ-fine partition $\mathcal{T} = \{([\bar{t}, t], \tau)\}$ of $[a, b]$ such that $([\sigma, t], \sigma) \in \mathcal{T}$. Then

$$\|u(T) - u(\sigma) - U(\sigma, T) + U(\sigma, \sigma)\|$$

$$\leq \sum_{\mathcal{T}} \|u(t) - u(\bar{t}) - U(\tau, t) + U(\tau, \bar{t})\| \leq \varepsilon.$$

Similarly if $a < \sigma - \delta(\sigma) \leq T < \sigma \leq b$ then

$$\|u(T) - u(\sigma) - U(\sigma, T) + U(\sigma, \sigma)\| \leq \varepsilon.$$

Hence (14.20) holds. The proof is complete. $\qquad\square$

14.18. Corollary. *$u(\sigma+)$ exists if and only if $U(\sigma, \sigma+)$ exists and then*

$$u(\sigma+) - u(\sigma) = U(\sigma, \sigma+) - U(\sigma, \sigma). \tag{14.21}$$

Let $a \leq S < T \leq b$ and let $u(S+)$ exist. The equation

$$u(T) - u(S) = u(T) - u(S+) + u(S+) - u(S)$$

can be written in the form

$$(\text{SKH}) \int_S^T D_t U(\tau, t) = (\text{SKH}) \int_{S+}^T D_t U(\tau, t) + (\text{SKH}) \int_S^{S+} D_t U(\tau, t) \tag{14.22}$$

where

$$(\text{SKH}) \int_S^{S+} D_t U(\tau, t) = U(S, S+) - U(S, S). \tag{14.23}$$

Similarly $u(\sigma-)$ exists if and only if $U(\sigma, \sigma-)$ exists and

$$u(\sigma) - u(\sigma-) = U(\sigma, \sigma) - U(\sigma, \sigma-). \tag{14.24}$$

Let $u(T-)$ exist. Then

$$(\text{SKH})\int_S^T D_t U(\tau, t) = (\text{SKH})\int_S^{T-} D_t U(\tau, t) + (\text{SKH})\int_{T-}^T D_t U(\tau, t) \tag{14.25}$$

where

$$(\text{SKH})\int_{T-}^T D_t U(\tau, t) = U(T, T) - U(T, T-). \tag{14.26}$$

In particular, u is continuous at σ if and only if $U(\sigma, .)$ is continuous at σ.

14.19. Lemma. *Let $[a, b] \subset \mathbb{R}$, let $\Phi : [a, b] \to \mathbb{R}$ be nondecreasing and let U fulfil*

$$\|U(\tau, t) - U(\tau, \bar{t})\| \le \Phi(t) - \Phi(\bar{t}) \quad \text{for } (\tau, t), (\tau, \bar{t}) \in \text{Dom}\, U.$$

Assume that U is SKH-integrable and that $u : [a, b] \to X$ is its primitive. Then

$$\|u(T) - u(S)\| \le \Phi(T) - \Phi(s) \quad \text{for } T, S \in [a, b]. \tag{14.27}$$

Proof. Let $a \le S < T \le b$. For $\varepsilon > 0$ there exists $\delta : [a, b] \to \mathbb{R}^+$ such that

$$\sum_{\mathcal{T}} \|u(t) - u(\bar{t}) - U(\tau, t) + U(\tau, \bar{t})\| \le \varepsilon,$$

$\mathcal{T}$ being a δ-fine partition of $[S, T]$. Hence

$$\|u(T) - u(S)\| \le \sum_{\mathcal{T}} \|u(t) - u(\bar{t})\|$$

$$\le \sum_{\mathcal{T}} \|U(\tau, t) - U(\tau, \bar{t})\| + \varepsilon \le \Phi(T) - \Phi(s) + \varepsilon$$

and (14.27) holds since ε is arbitrary positive. The proof is complete.   □

14.20. Theorem. *Let U be SKH-integrable on $[S, b]$ for $a < S < b$, $w \in X$ and*

$$\lim_{S \to a} \left(-(\text{SKH})\int_S^b D_t U(\tau, t) + w - U(a, S) + U(a, a) \right) = 0. \tag{14.28}$$

Then U is SKH-integrable on $[a,b]$ and

$$(\text{SKH}) \int_a^b D_t U(\tau,t) = w.$$

Proof. Let

$$u(S) = \begin{cases} -(\text{SKH}) \displaystyle\int_S^b D_t U(\tau,t) & \text{for } a < S \le b, \\ -w & \text{for } S = a. \end{cases} \tag{14.29}$$

Then

$$u(b) = 0, \quad u(b) - u(S) = (\text{SKH}) \int_S^b D_t U(\tau,t) \text{ for } a < S \le B$$

and u is an SKH-primitive of U on $[S,b]$ for $a < S < b$. Put $a_i = a + (b-a)\,2^{-i}$ for $i \in \mathbb{N}_0$. There exist $\delta_i : [a_i,b] \to \mathbb{R}^+$ for $i = 2,3,4,\ldots$ such that

$$\sum_A \|u(t) - u(\bar{t}) - U(\tau,t) + U(\tau,\bar{t})\| \le \varepsilon\,2^{-i-2} \tag{14.30}$$

for a δ_i-fine system A in $[a_{i+2}, b]$ (cf. Lemma 14.4). By (14.28) there exists ξ, $a < \xi < b$, such that

$$\left\| -U(a,S) + U(a,a) - (\text{SKH}) \int_S^b D_t U(\tau,t) + w \right\| \le \frac{\varepsilon}{2} \quad \text{for } a < S \le \xi.$$

Hence

$$\|u(S) - u(a) - U(a,S) + U(a,a)\| \le \frac{\varepsilon}{2} \tag{14.31}$$

since (cf. (14.29))

$$u(S) - u(a) - U(a,S) + U(a,a)$$
$$= -(\text{SKH}) \int_S^b D_t U(\tau,t) + w - U(a,S) + U(a,a).$$

There exists $\delta : [a,b] \to \mathbb{R}^+$ fulfilling

$$\delta(\tau) = \begin{cases} \xi & \text{for } \tau = a, \\ \min\{(b-a)\,2^{-i-2}, \delta_i(\tau)\} \\ \qquad \text{for } a_{i+1} < \tau \le a_i,\ i \in \mathbb{N}_0. \end{cases} \tag{14.32}$$

If $a < \tau \leq b$ then there exists a unique $i \in \mathbb{N}_0$ such that

$$
\left.
\begin{aligned}
&a_{i+1} < \tau \leq a_i, \quad a_{i+2} = a_{i+1} - \frac{b-a}{2^{i+2}} < \tau - \delta(\tau), \\
&[\tau - \delta(\tau), b] \subset [a_{i+2}, b].
\end{aligned}
\right\} \tag{14.33}
$$

Let $\mathcal{A} = \{([\bar{t}, t], \tau)\}$ be a δ-fine partition of $[a, b]$. (14.32) implies that there exists $([a, S], a) \in A$, $a < S \leq \xi$. Let

$$
\mathcal{A}_k = \{([\bar{t}, t], \tau) \in \mathcal{A}; a_{k+1} < \tau \leq a_k\} \quad \text{for} \quad k \in \mathbb{N}_0.
$$

If $([\bar{t}, t], \tau) \in \mathcal{A}_k$, then $\tau \in [a_{k+1}, a_k]$, $[\bar{t}, t] \subset [a_{k+2}, b]$ and by (14.30)

$$
\sum_{\mathcal{A}_k} \|u(t) - u(\bar{t}) - U(\tau, t) + u(\tau, \bar{t})\| \leq \varepsilon \, 2^{-k-2}. \tag{14.34}
$$

Moreover,

$$
\mathcal{A} = \{([a, S], a)\} \cup \bigcup_{k=0}^{\infty} \mathcal{A}_k.
$$

Hence (cf. (14.26), (14.24))

$$
\sum_{\mathcal{A}} \|u(t) - u(\bar{t}) - U(\tau, t) + U(\tau, \bar{t})\|
$$

$$
= \|u(S) - u(a) - U(S, a) + U(a, a)\|
$$

$$
+ \sum_{k=0}^{\infty} \sum_{\mathcal{A}_k} \|u(t) - u(\bar{t}) - U(\tau, t) + U(\tau, \bar{t})\|
$$

which implies that U is SKH-integrable on $[a, b]$, that u is its primitive and

$$
(\text{SKH}) \int_a^b D_t U(\tau, t) = u(b) - u(a) = w.
$$

$\qquad\qquad\qquad\qquad\qquad\qquad\qquad\qquad\qquad\qquad\qquad\qquad\qquad\quad\square$

14.21. Remark. Theorem 14.20 is an extension of Hake's theorem (cf. [Schwabik, Ye (2005)], Theorem 3.4.5 and Remark) to the integration of functions of a pair of coupled variables.

The next theorem is an analogue of Theorem 14.20 and can be proved in a similar way.

14.22. Theorem. *Let U be SKH-integrable on $[a, T]$ for $a < T < b$, $x \in X$ and let*

$$
\lim_{T \to b} \left(-(\text{SKH}) \int_a^T D_t U(\tau, t) + w + U(b, t) - U(b, b) \right) = 0.
$$

Then U is SKH-integrable on $[a, b]$ and

$$(\text{SKH}) \int_a^b D_t U(\tau, t) = w.$$

14.23. Remark. Let $f : [a, b] \to X$ and

$$U(\tau, t) = f(\tau) t \quad \text{for } (\tau, t) \in [a, b]^2.$$

Then f is called SKH-*integrable* on $[a, b]$ if U is SKH-integrable on $[a, b]$ and $(\text{SKH}) \int_a^b f(t) \, dt$ is written instead of $(\text{SKH}) \int_a^b D_t U(\tau, t)$. Furthermore, u is called an SKH-*primitive* of f if u is a primitive of U. Therefore f is SKH-integrable on $[S, T]$ and u is its primitive if and only if for every $\varepsilon > 0$ there exists $\delta : [a, b] \to \mathbb{R}^+$ such that

$$\sum_{\mathcal{A}} \| f(\tau)(t - \bar{t}) - u(t) + u(\bar{t}) \| \leq \varepsilon \tag{14.35}$$

for every δ-fine partition $\mathcal{A} = \{([\bar{t}, t], \tau)\}$ of $[S, T]$. Moreover,

$$(\text{SKH}) \int_a^b f(t) \, dt = u(b) - u(a).$$

14.24. Lemma. *Let $Q \subset [a, b]$, $|Q| = 0$, $h : Q \to \mathbb{R}_0^+$, $\varepsilon > 0$. Then there exists $\xi : Q \to \mathbb{R}^+$ such that*

$$\sum_{\mathcal{A}} h(\tau) \, (t - \bar{t}) \leq \varepsilon$$

if $A = \{([\bar{t}, t], \tau)\}$ is a ξ-fine Q-anchored system in $[a, b]$.

Hint. Let $Q_0 = \emptyset$, $Q_j = \{t \in Q ; h(t) \leq 2^j\}$, $j \in \mathbb{N}$. There exist open sets $G_j \subset \mathbb{R}$, $j \in \mathbb{N}$, such that $Q_j \subset G_j$, $|G_j| \leq \varepsilon \, 2^{-2j}$. Let $(t - \xi(t), t + \xi(t)) \subset G_j$ for $t \in Q_j \setminus Q_{j-1}$. Then ξ does the job.

14.25. Theorem. *Let $f : [a, b] \to X$, $u : [a, b] \to X$. Then*

$$f \quad \text{is SKH -integrable and } u \text{ is its primitive on } [a, b] \tag{14.36}$$

if and only if

$$\left. \begin{array}{l} \text{there exists } Q \subset [a, b] \quad \text{such that } |Q| = 0, \\[2mm] \dfrac{d}{dt} u(t) = f(t) \quad \text{for } t \in [a, b] \setminus Q, \end{array} \right\} \tag{14.37}$$

and

$$\left.\begin{array}{l} \textit{for each } \varepsilon > 0 \textit{ there exists } \delta_4 : [a,b] \to \mathbb{R}^+ \textit{ such that} \\[2mm] \displaystyle\sum_{\mathcal{A}} \|u(t) - u(\bar{t})\| \leq \varepsilon \quad \textit{for each} \\[2mm] \delta_4\textit{-fine } Q\textit{-anchored system } A = \{([\bar{t}, t], \tau)\} \textit{ in } [a, b]. \end{array}\right\} \tag{14.38}$$

Proof. Let (14.37) and (14.38) hold. Then for every $\varepsilon > 0$ there exists $\delta_1 : [a, b] \to \mathbb{R}^+$ such that

$$\sum_{\substack{([\bar{t},t],\tau) \in \mathcal{A}, \\ \tau \in [a,b] \setminus Q}} \|u(t) - u(\bar{t}) - f(\tau)(t - \bar{t})\| + \sum_{\substack{([\bar{t},t],\tau) \in \mathcal{A}, \\ \tau \in Q}} \|u(t) - u(\bar{t})\| \leq \frac{\varepsilon}{2}$$

for every δ_1-fine partition $\mathcal{A} = \{([\bar{t}, t], \tau)\}$ of $[a, b]$. Moreover (cf. Lemma 14.24), there exists $\delta_2 : [a, b] \to \mathbb{R}^+$ such that

$$\sum_{\mathcal{C}} \|f(\tau)\|(t - \bar{t}) \leq \frac{\varepsilon}{2}$$

for every δ_2-fine Q-anchored system $\mathcal{C} = \{([\bar{t}, t], \tau)\}$ in $[a, b]$. Hence

$$\sum_{\mathcal{A}} \|f(\tau)(t - \bar{t}) - u(t) + u(\bar{t})\| \leq \varepsilon$$

for a δ-fine partition $\mathcal{A} = \{([t, \bar{t}], \tau)\}$ of $[a, b]$ where $\delta(\tau) = \delta_1(\tau)$ if $\tau \in [a, b] \setminus Q$ and $\delta(\tau) = \delta_2(\tau)$ if $\tau \in Q$.

On the other hand, let (14.36) hold. By Theorem 7.4.2 in [Schwabik, Ye (2005)] there exists Q such that (14.37) holds. Let $\varepsilon > 0$. By Lemma 14.24 there exists $\delta_4 : [a, b] \to \mathbb{R}^+$ which fulfils

$$\sum_{\mathcal{A}} \|f(\tau)\|(t - \bar{t}) \leq \tfrac{1}{2}\varepsilon \tag{14.39}$$

for every δ_4-fine Q-anchored system $A = \{([\bar{t}, t], \tau)\}$ in $[a, b]$. (14.38) is correct. $\square$

14.26. Remark. The concept of a real-valued ACG^* function was extended to Banach-valued functions in a natural way (see [Schwabik, Ye (2005)], Definition 7.1.5 (d)). By [Schwabik, Ye (2005)], Theorem 7.4.5, the function $f : [a, b] \to X$ is SKH-integrable on $[a, b]$ and $v : [a, b] \to X$ is its primitive if and only if v is continuous and ACG^* on $[a, b]$ such that $\dfrac{dv}{dt}(t) = f(t)$ almost everywhere in $[a, b]$.

Observe that the set of primitives is the set of continuous ACG^* functions.

14.27. Remark. If $\varphi : [a, b] \to \mathbb{R}$ is Lebesgue integrable, $c \in [a, b]$,

$$u(t) = (\mathrm{L}) \int_c^t \varphi(s)\, \mathrm{d}s \quad \text{(Lebesgue integral)} \quad \text{for } t \in [a, b]$$

then (14.37), (14.38) are fulfilled and, by Theorem 14.25, φ is SKH-integrable on $[a, b]$ and

$$(\mathrm{SKH}) \int_c^t \varphi(s)\, \mathrm{d}s = (\mathrm{L}) \int_c^t \varphi(s)\, \mathrm{d}s, \quad t \in [a, b].$$

Chapter 15

Generalized differential equations: Strong Kurzweil Henstock solutions

15.1. Notation. Let

$$\operatorname{Dom} G \subset X \times \mathbb{R}^2, \quad G : \operatorname{Dom} G \to X, \quad [a, b] \subset \mathbb{R},$$
$$\delta_0 : [a, b] \to \mathbb{R}^+, \quad u : [a, b] \to X,$$
$$(u(\tau), \tau, t) \in \operatorname{Dom} G \quad \text{for } \tau, t \in [a, b], \quad \tau - \delta_0(\tau) \le t \le \tau + \delta_0(\tau).$$

15.2. Definition. u is an SKH-*solution* of

$$\frac{\mathrm{d}}{\mathrm{d}t} x = \mathrm{D}_t G(x, \tau, t) \tag{15.1}$$

on $[a, b]$ if

$$u(T) - u(S) = (\mathrm{SKH}) \int_S^T \mathrm{D}_t G(u(\tau), \tau, t) \quad \text{for } [S, T] \subset [a, b]. \tag{15.2}$$

15.3. Theorem. *The following three assertions are equivalent:*

(i)   *u is an SKH-solution of* (15.1),

(ii)   *U is SKH-integrable on $[a, b]$ and u is its primitive,*

$$U(\tau, t) = G(u(\tau), \tau, t),$$

(iii)   *for every $\varepsilon > 0$ there exists a $\delta : [a, b] \to \mathbb{R}^+$ such that*

$$\sum_{\mathcal{A}} \| u(t) - u(\bar{t}) - G(u(\tau), \tau, t) + G(u(\tau), \tau, \bar{t}) \| \le \varepsilon,$$

$$\mathcal{A} = \{([\bar{t}, t], \tau)\} \quad \text{being any } \delta\text{-fine partition of } [a, b].$$

Proof. The theorem is a consequence of Theorem 14.15.   □

15.4. Theorem. *If u is an SR-solution of (15.1) then it is an SKH-solution of (15.1).*

Proof.   This is a consequence of Definitions 15.2 and 6.2 (cf. also Remark 14.12).   □

15.5. Lemma. *Let u be an SKH-solution of (15.1) on $[a,b]$, $[S,T] \subset [a,b]$. Then u is an SKH-solution of (15.1) on $[S,T]$.*

Proof.   This is a consequence of Corollary 14.8.   □

15.6. Lemma. *Let $a < c < b$ and let u be an SKH-solution of (15.1) on $[a,c]$ and on $[c,b]$. Then u is an SKH-solution of (15.1) on $[a,b]$.*

Proof. u is an SKH-solution of (15.1) on $[a,b]$ by Lemma 14.13.   □

15.7. Lemma. *Let u be an SKH-solution of (15.1)) on $[a,b]$, $c \in [a,b]$. Assume that $G(u(c),c,\cdot)$ is continuous at c. Then u is continuous at c.*

Proof. u is continuous at c by Corollary 14.18.   □

15.8. Theorem. *Let $u:[a,b] \to X$ be an SKH-solution of (15.1) on $[c,b]$ for $a < c < b$. Assume that*

$$\left. \begin{array}{r} u(S) - u(a) - G(u(a),a,S) + G(u(a),a,a) \to 0 \\[2mm] \text{for } \ S \to a, \ S > a. \end{array} \right\} \qquad (15.3)$$

Then

$$u \quad \text{is an SKH-solution of (15.1) on } \ [a,b]. \qquad (15.4)$$

Proof.   Put

$$U(\tau,t) = G(u(\tau),\tau,t) \quad \text{for } \ \tau,t \in [a,b], \ |t-\tau| \le \delta_0(\tau).$$

Then

$$u(c) = u(b) - (\text{SKH}) \int_c^b D_t U(\tau,t),$$

since u is an SKH-solution of (15.1) on $[c,b]$. Put $w = u(b) - u(a)$. Then

$$-(\text{SKH}) \int_c^b U(\tau,t) + w - U(a,c) + U(a,a)$$

$$- u(c) \quad u(a) - G(u(a),a,c) + G(u(a),a,a)$$

Hence (14.28) is fulfilled by (15.3).

Theorem 14.20 implies that $(\mathrm{SKH})\displaystyle\int_a^b D_t G(u(\tau),\tau,t)$ exists and equals $u(b) - u(a)$. Theorem 15.8 is correct. □

15.9. Remark. Theorem 15.8 is an extension of Hake's theorem to SKH-solutions of (15.1), cf. Remark 14.21.

15.10. Lemma. *Let* $\widehat{R} > 0$. *Assume that the functions*

$$G : B(\widehat{R}) \times [-b, -a]^2 \to X, \quad H : B(\widehat{R}) \times [a, b]^2 \to X,$$
$$u : [-b, -a] \to B(\widehat{R}), \qquad w : [a, b] \to B(\widehat{R}),$$
$$\delta : [-b, -a] \to \mathbb{R}^+, \qquad \rho : [a, b] \to \mathbb{R}^+$$

are coupled by the relations

$$G(x, \tau, t) = H(x, \sigma, s), \quad u(\tau) = w(\sigma), \quad \delta(\tau) = \rho(\sigma),$$
$$\tau = -\sigma, \quad t = -s \quad \text{for } x \in B(\widehat{R}), \ \tau, t \in [a, b].$$

Then u *is an* SKH-*solution of* (15.1) *on* $[-b, -a]$ *if and only if* w *is an* SKH-*solution of*

$$\frac{\mathrm{d}}{\mathrm{d}s} x = D_s H(x, \sigma, s) \tag{15.5}$$

on $[a, b]$.

Proof.   Let the sets

$$\{t_0, \tau_1, t_1, \tau_2 \ldots, \tau_k, t_k\} \subset [-b, -a], \quad \{s_0, \sigma_1, s_1, \sigma_2, \ldots, \sigma_k, s_k\} \subset [a, b]$$

be coupled by

$$s_i = -t_i \ \text{ for } i = 0, 1, \ldots, k, \quad \sigma_i = -\tau_i \ \text{ for } i = 1, 2, \ldots, k.$$

Then $\{([t_{i-1}, t_i], \tau_i) ; i = 1, 2, \ldots, k\}$ is a δ-fine partition of $[-b, -a]$ if and only if $\{([s_i, s_{i-1}], \sigma_i) ; i = 1, 2, \ldots, k\}$ is a ρ-fine partition of $[a, b]$. Moreover,

$$\|u(t_i) - u(t_{i-1}) - G(u(\tau_i), \tau_i, t_i) + G(u(\tau_i), \tau_i, t_{i-1})\|$$
$$= \|w(s_{i-1}) - w(s_i) - H(w(\sigma_i), \sigma_i, s_{i-1}) + H(w(\sigma_i), \sigma_i, s_i)\|$$

and the lemma is correct. □

15.11. Remark. Let $\mathrm{Dom}\, g \subset X \times \mathbb{R}$, $u : [a, b] \to X$ and let $(u(t), t) \in \mathrm{Dom}\, g$ for $t \in [a, b]$. Then u is called an SKH-solution of

$$\frac{\mathrm{d}}{\mathrm{d}t} x = g(x, t) \tag{15.6}$$

on $[a, b]$ if (cf. Remark 14.23)

$$u(T) - u(S) = (\mathrm{SKH}) \int_S^T g(u(t), t)\, \mathrm{d}t \text{ for } [S, T] \subset [a, b]. \tag{15.7}$$

15.12. Theorem. *u is an (SKH)-solution of (15.6) on $[a, b]$ if and only if*

$$\left.\begin{array}{l} \text{there exists } Q \subset [a, b] \quad \text{such that } |Q| = 0, \\[2mm] \dfrac{\mathrm{d}}{\mathrm{d}t} u(t) = g(u(t), t) \quad \text{for } t \in [a, b] \setminus Q \end{array}\right\} \tag{15.8}$$

and

$$\left.\begin{array}{l} \text{for each } \varepsilon > 0 \text{ there exists } \delta_4 : [a, b] \to \mathbb{R}^+ \text{ such that} \\[2mm] \displaystyle\sum_A \|u(t) - u(\bar{t})\| \leq \varepsilon \quad \text{for each } \delta_4\text{-fine} \\[2mm] Q\text{-anchored system } A = \{([\bar{t}, t], \tau)\} \text{ in } [a, b]. \end{array}\right\} \tag{15.9}$$

This is a consequence of Theorem 14.25.

Chapter 16

Uniqueness

16.1. Notation. Let a, b, c fulfil (12.2). Assume that

$$G : B(8R) \times [a, c]^2 \to X \quad \text{fulfils (8.2)–(8.7)} . \tag{16.1}$$

16.2. Theorem. *Let $[S, T] \subset [a, b]$ and let $w : [S, T] \to B(4R)$ be an SKH-solution of (8.8). Then there exists a unique $z^* \in B(5R)$ such that*

$$w(s) = z^* W(a, s) \quad \text{for} \quad s \in [S, T] . \tag{16.2}$$

Proof. By Theorem 12.10 there exists a $z^* \in B(5R)$ such that

$$w(S) = z^* W(a, S) . \tag{16.3}$$

z^* is unique by (12.8) and (12.1).

Let $0 < \varepsilon < \frac{1}{3}$, $S < s \leq T$. By Theorem 15.3 there exists $\delta_1 : [S, T] \to \mathbb{R}^+$ such that

$$\sum_{\mathcal{A}} \| w(t) - w(\bar{t}) - G(w(\tau), \tau, t) + G(w(\tau), \tau, \bar{t}) \| \leq \varepsilon \tag{16.4}$$

for every δ_1-fine partition $\mathcal{A} = \{([\bar{t}, t], \tau)\}$ of $[S, s]$. Moreover, by Lemma 12.7 there exists η, $0 < \eta < s - S$ such that

$$\left. \begin{aligned} \| x\, W(a, t) - x\, W(a, \bar{t}) \\ - G(x\, W(a, \tau), \tau, t) + G(x\, W(a, \tau), \tau, \bar{t}) \| \\ \leq \varepsilon\, (t - \bar{t})\, (b - a)^{-1} \end{aligned} \right\} \tag{16.5}$$

if $a \leq \bar{t} \leq \tau \leq t \leq s$, $[\bar{t}, t] \subset [\tau - \frac{\eta}{2}, \tau + \frac{\eta}{2}]$.

Put

$$\delta_2(\tau) = \min\left\{\delta_1(\tau), \tfrac{\eta}{2}, \tfrac{1}{2}(\tau - S), \tfrac{1}{2}(s - \tau)\right\}$$
$$\text{for } S < \tau < s,$$

$$\delta_2(S) = \min\left\{\delta_1(S), \tfrac{\eta}{2}\right\},$$

$$\delta_2(s) = \min\left\{\delta_1(s), \tfrac{\eta}{2}\right\}.$$

(16.6)

By Lemma 14.4 there exists a δ_2-fine partition $C = \{([t_{i-1}, t_i], \tau_i);$ $i = 1, 2, \ldots, k\}$ of $[S, s]$, i.e.

$$S = t_0 \le \tau_1 \le t_1 \le \tau_2 \le \ldots \le t_{k-1} \le \tau_k \le t_k = s, \ t_0 < t_1 < \cdots < t_{k-1} < t_k.$$

(16.6) implies that $S < \tau - \delta_2(\tau) < \tau + \delta_2(\tau) < s$ if $S < \tau < T$ so that

$$S = t_0 = \tau_1 < t_1, \ t_{k-1} < \tau_k = t_k = s. \tag{16.7}$$

By Theorem 12.10 there exist $z_i \in B(5R)$ such that

$$z_i \, W(a, \tau_i) = w(\tau_i), \quad i = 1, 2, \ldots, k.$$

Observe that

$$z^* = z_1. \tag{16.8}$$

Put

$$u_i(t) = z_i \, W(a, t) \quad \text{for } t \in [a, b], \ i = 1, 2, \ldots, k. \tag{16.9}$$

By Theorem 12.10, u_i are SR-solutions of (8.8) fulfilling

$$u_i(\tau_i) = w(\tau_i), \quad i = 1, 2, \ldots, k. \tag{16.10}$$

Partition C is modified to

$$\mathcal{D} = \{([t_{i-1}, \tau_i], \tau_i), ([\tau_i, t_i], \tau_i), \quad i = 1, 2, \ldots, k\}$$

which again is a δ_2-fine partition of $[S, s]$. Put

$$\Lambda_i = \|w(t_i) - w(\tau_i) - G(w(\tau_i), \tau_i, t_i)$$
$$+ G(w(\tau_i), \tau_i, \tau_i)\|, \quad i = 1, 2, \ldots, k - 1,$$

(16.11)

$$\lambda_i = \|w(t_i) - w(\tau_{i+1}) - G(w(\tau_{i+1}), \tau_{i+1}, t_i)$$
$$+ G(w(\tau_{i+1}), \tau_{i+1}, \tau_{i+1})\|, \ i = 1, 2, \ldots, k - 1.$$

(16.12)

Then

$$\sum_{i=1}^{k-1}(\Lambda_i + \lambda_i) \leq \varepsilon \tag{16.13}$$

since D is δ_2-fine. Moreover, (cf. (16.5))

$$\left.\begin{aligned}&\|u_i(t_i) - u_i(\tau_i) - G(u_i(\tau_i), \tau_i, t_i) + G(u_i(\tau_i), \tau_i)\|\\ &\qquad\leq \varepsilon \frac{t_i - \tau_i}{b-a}, \quad i = 1, 2, \ldots, k-1,\end{aligned}\right\} \tag{16.14}$$

$$\left.\begin{aligned}&\|u_{i+1}(t_i) - u_{i+1}(\tau_{i+1})\\ &\qquad - G(u_{i+1}(\tau_{i+1}), \tau_{i+1}, t_i) + G(u_{i+1}(\tau_{i+1}), \tau_{i+1}, \tau_{i+1})\|\\ &\qquad\leq \varepsilon \frac{\tau_{i+1} - t_i}{b-a}, \quad i = 1, 2, \ldots, k-1.\end{aligned}\right\} \tag{16.15}$$

(16.11) and (16.14) imply that

$$\|w(t_i) - u_i(t_i)\| \leq \Lambda_i + \varepsilon \frac{t_i - \tau_i}{b-a}, \quad i = 1, 2, \ldots, k-1 \tag{16.16}$$

since $u_i(\tau_i) = w(\tau_i)$. Similarly (16.12) and (16.15) imply that

$$\|w(t_i) - u_{i+1}(t_i)\| \leq \lambda_i + \varepsilon \frac{\tau_{i+1} - t_i}{b-a}, \quad i = 1, 2, \ldots, k-1 \tag{16.17}$$

since $u_{i+1}(\tau_{i+1}) = w(\tau_{i+1})$. By (16.13), (16.16), (16.17)

$$\|u_i(t_i) - u_{i+1}(t_i)\| \leq \varepsilon + \varepsilon \frac{\tau_{i+1} - \tau_i}{b-a} \leq \frac{2}{3}, \quad i = 1, 2, \ldots, k-1 \tag{16.18}$$

since $\varepsilon < \frac{1}{3}$. By (12.10), where

$$S = a, \ t = t_i, \ v = u_i(a) = z_i, \ \bar{v} = u_{i+1}(a) = z_{i+1},$$
$$v\,W(a, t_i) = u_i(t_i), \quad \bar{v}\,W(a, t_i) = u_{i+1}(t_i),$$

we have

$$\left.\begin{aligned}\|u_i(a) - u_{i+1}(a)\| &\leq \frac{\|u(t_i) - u_{i+1}(t_i)\|}{1 - \psi_2(t-S) - \mathcal{B}(2(t-S))}\\ &\leq \frac{3}{2}\left(\Lambda_i + \lambda_i + \varepsilon \frac{\tau_{i+1} - \tau_i}{b-a}\right), \quad i = 1, 2, \ldots, k-1,\end{aligned}\right\} \tag{16.19}$$

and

$$\|u_1(a) - u_k(a)\| \leq \frac{3}{2}\,2\,\varepsilon = 3\,\varepsilon \tag{16.20}$$

(cf. (16.13), (12.1)), (12.2)).

Finally,

$$u_1(a) = z_1 = z^* \quad (\text{cf. (16.8)}) \,, u_k(a) = z_k, \; \|z_k - z^*\| \le 3\,\varepsilon,$$

$$z_k \, W(a, s) = u_k(s) = w(s)$$

and (cf. (12.8))

$$\|z^* \, W(a, s) - z_k \, W(a, s)\| \le \frac{4}{3} \, \|z^* - z_k\| \le 4\varepsilon,$$

i.e.,

$$\|z^* \, W(a, s) - w(s)\| \le 4\varepsilon,$$

which proves (16.2). $\qquad\qquad\qquad\qquad\qquad\qquad\qquad\qquad\qquad\qquad$ $\square$

16.3. Remark. Every (local) SKH-solution u of (8.8) is by Theorem 16.2 an SR-solution of (8.8) and is uniquely determined by the Cauchy condition $u(s) = y$.

Chapter 17

Differential equations in classical form

17.1. Notation. $\operatorname{Dom} g \subset X \times \mathbb{R}$, $g : \operatorname{Dom} g \to X$, $[a,b] \subset \mathbb{R}$, $u : [a,b] \to X$, $(u(\tau), \tau) \in \operatorname{Dom} g$ for $\tau \in [a,b]$.

17.2. Definition. u is an SKH-*solution* of

$$\dot{x} = g(x,t) \tag{17.1}$$

if

$$u(T) - u(S) = (\text{SKH}) \int_S^T g(u(t),t)\, \mathrm{d}t \quad \text{for } [S,T] \subset [a,b]. \tag{17.2}$$

17.3. Theorem. *The seven conditions below are equivalent.*

  (i) u *is an* SKH-*solution of* (17.1),

  (ii) $u(T) - u(S) = (\text{SKH}) \displaystyle\int_S^T D_t U(\tau,t)$, *where*

$$U(\tau,t) = g(u(\tau),\tau)\, t \text{ for } \tau,\, t \in [a,b],\ a \le S < T \le b,$$

  (iii) U *is* SKH-*integrable on* $[a,b]$, u *being its primitive,*

  (iv) *for every* $\varepsilon > 0$ *there exists a* $\delta : [a,b] \to \mathbb{R}^+$ *such that*

$$\sum_A \| u(t) - u(\bar{t}) - g(u(\tau),\tau)\,(t - \bar{t}) \| \le \varepsilon,$$

  $\mathcal{A}$ *being any* δ-*fine partition of* $[a,b]$,

  (v) *there exists* $Q \subset [a,b]$, $|Q| = 0$ *such that*

$$\frac{\mathrm{d}u}{\mathrm{d}t}(t) = g(u(t),t) \quad \text{for } t \in [a,b] \setminus Q \tag{17.3}$$

105

and

$$\left.\begin{array}{l} \textit{for every } \varepsilon > 0 \quad \textit{there exists a } \delta_1 : [a,b] \to \mathbb{R}^+ \\[4pt] \textit{such that} \\[4pt] \qquad \sum_A \|u(t) - u(\bar{t})\| \leq \varepsilon \\[4pt] \textit{if } A = \{([\bar{t}, t], \tau]\} \\[4pt] \textit{is a } \delta_1 - \textit{fine } Q - \textit{anchored system in } [a,b], \end{array}\right\} \tag{17.4}$$

(vi) $g(u(t), t)$ *is SKH-integrable on* $[a,b]$ *and* u *is its primitive,*

(vii) u *is continuous and* ACG^* *on* $[a,b]$ *and*

$$\frac{\mathrm{d}u}{\mathrm{d}t}(t) = g(u(t), t)$$

almost everywhere.

Proof.   (i) and (ii) are equivalent since $U(\tau, t) = g(\tau) t$. (ii)–(iv) are equivalent by Theorem 15.3. (iii) and (v) are equivalent by Theorem 14.25. (ii) and (vi) are equivalent since

$$(\text{SKH}) \int_S^T g(u(s), s) \, \mathrm{d}s = (\text{SKH}) \int_S^T \mathrm{D}_t U(\tau, t) \quad \text{for } [S, T] \subset [a, b]$$

by Remark 14.23. (vi) and (vii) are equivalent by Remark 14.26.   □

17.4 . Theorem. *Assume that* $a < T \leq b$, *that* $u : [a,b] \to X$ *is an SKH-solution of* (17.1) *on* $[S, T]$ *for* $a < S < T$ *and that*

$$u(a) = \lim_{S \to a} u(S). \tag{17.5}$$

Then u *is an SKH-solution of* (17.1) *on* $[a, T]$.

Proof.   Put $\operatorname{Dom} G = (\operatorname{Dom} g) \times \mathbb{R}$, $G(x, \tau, t) = g(x, \tau) t$.

By Theorem 17.3 (i) and (ii), u is an SKH-solution of the GODE

$$\frac{\mathrm{d}}{\mathrm{d}t} x = \mathrm{D}_t G(x, \tau, t) \tag{17.6}$$

on $[S, T]$ for $a < S < T$. Then condition (15.3) is fulfilled since

$$G(u(a), a, c) \to G(u(a), a, a) \quad \text{for } c \to a$$

and (17.5) holds. By Theorem 15.8 u is an SKH-solution of (17.6) on $[a, T]$. By Theorem 17.3 (i) and (ii), u is an SKH-solution of (17.1) on $[a, T]$. $\square$

17.5. Lemma. *Let* $u : [a, b] \to X$, $a < c < b$ *and let the restrictions* $u|_{[a,c]}$, $u|_{[c,b]}$ *be* SKH-*solutions of* (17.1) *on* $[a, c]$ *and on* $[c, b]$ *respectively. Then* u *is an* SKH-*solution of* (17.1) *on* $[a, b]$.

Proof. u is an SKH-solution of (17.1) by Lemma 15.6 and Theorem 17.3 (i), (ii). $\square$

17.6. Theorem. *Let* $\operatorname{Dom} G \subset X \times \mathbb{R}^2$, $G : \operatorname{Dom} G \to X$, $\delta_0 : [a, b] \to \mathbb{R}^+$, $u : [a, b] \to X$, $\operatorname{Dom} g \subset X \times \mathbb{R}$, $g : \operatorname{Dom} g \to X$,

$$(u(\tau), \tau, t) \in \operatorname{Dom} G \quad for \ \tau, t \in [a, b], \ |t - \tau| \leq \delta_0(\tau), \tag{17.7}$$

$$(u(\tau), \tau) \in \operatorname{Dom} g \quad for \ \tau \in [a, b], \tag{17.8}$$

$(u(\tau), \tau) \in \operatorname{Dom} g$ *for* $\tau \in [a, b]$, $Q \subset [a, b]$, $|Q| = 0$.

Assume that

$$\frac{\partial}{\partial t} G(u(\tau), \tau, t)|_{t=\tau} = g(u(\tau), \tau) \quad for \ \tau \in [a, b] \setminus Q, \tag{17.9}$$

and for every $\varepsilon > 0$ *there exists* $\delta : [a, b] \to \mathbb{R}^+$ *such that*

$$\sum_{\mathcal{A}} \| G(u(\tau), \tau, t) - G(u(\tau), \tau, \bar{t}) \| \leq \varepsilon, \tag{17.10}$$

$\mathcal{A} = \{([\bar{t}, t], \tau)\}$ *being a* δ-*fine* Q-*anchored system in* $[a, b]$.

Then u *is an* SKH-*solution of* (17.1) *if and only if* u *is an* SKH-*solution of the GODE*

$$\frac{\mathrm{d}}{\mathrm{d}t} x = \mathrm{D}_t G(x, \tau, t). \tag{17.11}$$

Proof. Put

$$W(\tau, t) = g(u(\tau), \tau) t - G(u(\tau), \tau, t) \tag{17.12}$$

for $\tau, t \in [a, b]$, $|t - \tau| \leq \delta_0(\tau)$. Let $\mathcal{A} = \{([\bar{t}, t], \tau)\}$ be a δ_0-fine partition of $[a, b]$. Then

$$\sum_{\mathcal{A}} \| W(\tau, t) - W(\tau, \bar{t}) \| \leq \Theta_1 + \Theta_2 + \Theta_3 \tag{17.13}$$

where

$$\Theta_1 = \sum_{\mathcal{A};\tau \notin Q} \|g(u(\tau),\tau)(t-\bar{t}) - G(u(\tau),\tau,t) + G(u(\tau),\tau,\bar{t})\|\,,$$

$$\Theta_2 = \sum_{\tau \in Q} \|g(u(\tau),\tau)\,(t-\bar{t})\|\,,$$

$$\Theta_3 = \sum_{\tau \in Q} \|G(u(\tau),\tau,t) - G(u(\tau),\tau,\bar{t})\|\,.$$

Let $\varepsilon > 0$. By (17.9) there exists $\delta_1 : [a,b] \to \mathbb{R}^+$ such that

$$\|G(u(\tau),\tau,t) - G(u(\tau),\tau,\bar{t}) - g(u(\tau),\tau)\,(t-\bar{t})\| \le \varepsilon \frac{t-\bar{t}}{b-a}$$

if $\tau \in [a,b] \setminus Q$, $t,\bar{t} \in [a,b]$, $\tau - \delta_1(\tau) \le \bar{t} \le \tau \le t \le \tau + \delta_1(\tau)$. Hence

$$\|\Theta_1\| \le \varepsilon \quad \text{if} \quad \mathcal{A} \text{ is } \delta_1 - \text{fine}. \tag{17.14}$$

By Lemma 14.24 there exists $\delta_2 : [a,b] \to \mathbb{R}^+$ such that

$$\|\Theta_2\| \le \varepsilon \quad \text{if} \quad \mathcal{A} \text{ is } \delta_2 - \text{fine}. \tag{17.15}$$

By (17.10) there exists $\delta_3 : [a,b] \to \mathbb{R}^+$ such that

$$\|\Theta_3\| \le \varepsilon \quad \text{if} \quad \mathcal{A} \text{ is } \delta_3 - \text{fine}. \tag{17.16}$$

Let $\delta_4(\tau) = \min\{\delta_1(\tau),\delta_2(\tau),\delta_3(\tau)\}$ for $\tau \in [a,b]$. By (17.13)–(17.16)

$$\sum_{\mathcal{A}} \|W(\tau,t) - W(\tau,\bar{t})\| \le 3\varepsilon \quad \text{if} \quad \mathcal{A} \text{ is } \delta_4 - \text{fine}.$$

Hence

$$(\text{SKH}) \int_a^b D_t W(\tau,t) = 0$$

and Theorem 17.6 is correct. $\qquad\qquad\qquad\qquad\qquad\qquad\qquad\qquad\square$

The next lemma is a consequence of Lemma 15.10.

17.7. Lemma. *Let $\widehat{R} > 0$. Assume that functions*

$$g : B(\widehat{R}) \times [-b,-a] \to X, \quad f : B(\widehat{R}) \times [a,b] \to X\,,$$

$$u : [-b,-a] \to B(\widehat{R}), \qquad w : [a,b] \to B(\widehat{R})\,,$$

$$\delta : [-b,-a] \to \mathbb{R}^+, \qquad \widehat{\delta} : [a,b] \to \mathbb{R}^+$$

are coupled by the relations

$$g(x, \tau) = -f(x, \sigma), \quad u(\tau) = w(\sigma),$$
$$\delta(\tau) = \widehat{\delta}(\sigma) \quad \text{for} \ \ x \in B(\widehat{R}), \ \sigma = -\tau, \ \tau \in [a, b] \,.$$

Then u is an SKH-solution of (17.1) on $[-b, -a]$ if and only if w is an SKH-solution of

$$\dot{x} = f(x, t) \tag{17.17}$$

on $[a, b]$.

Chapter 18

On a class of differential equations in classical form

18.1. Notation. Let α, β, η, ν, $\widehat{R} \in \mathbb{R}^+$,

$$h : B(\widehat{R}) \times (\mathbb{R} \setminus \{0\}) \to X, \quad H : B(\widehat{R}) \times (\mathbb{R} \setminus \{0\}) \to X.$$

Assume that

$$1 \leq \alpha < 1 + \tfrac{1}{2}\beta, \quad \eta = \min\{1, 2 + \beta - 2\alpha\}, \tag{18.1}$$

$$h, \mathrm{D}_1 H, \mathrm{D}_2 H \text{ are continuous (cf. Notation 3.1)}, \tag{18.2}$$

$$\mathrm{D}_2 H(x, t) = \frac{\partial}{\partial t} H(x, t) = h(x, t) \quad \text{for } x \in B(\widehat{R}), \, t \in \mathbb{R} \setminus \{0\}, \tag{18.3}$$

$$\left. \begin{array}{l} \|h(x, t)\| \leq \nu, \; \|\Delta_v h(x, t)\| \leq \|v\| \, \nu \\[2mm] \qquad\qquad \text{for } x, x + v \in B(\widehat{R}), \, t \in \mathbb{R} \setminus \{0\}, \end{array} \right\} \tag{18.4}$$

$$\left. \begin{array}{l} \|H(x, t)\| \leq \nu, \; \|\Delta_v H(x, t)\| \leq \|v\| \, \nu, \; \|\Delta_v \mathrm{D}_1 H(x, t)\| \leq \|v\| \, \nu \\[2mm] \qquad\qquad \text{for } x, x + v \in B(\widehat{R}), \, t \in \mathbb{R} \setminus \{0\}, \end{array} \right\} \tag{18.5}$$

$$g(x, t) = \begin{cases} t^{-\alpha} h(x, t^{-\beta}) & \text{if } x \in B(\widehat{R}), \, 0 < t \leq 1, \\ 0 & \text{if } x \in B(\widehat{R}), \\ |t|^{-\alpha} h(x, -|t|^{-\beta}) & \text{if } x \in B(\widehat{R}), \, -1 \leq t < 0. \end{cases} \tag{18.6}$$

18.2. Remark. In this chapter couples $(\beta, \alpha) \in (\mathbb{R}^+)^2$ are found such that SKH-solutions of

$$\dot{x} = g(x, t) \tag{18.7}$$

exist on intervals $[-\mu, \mu]$, $\mu > 0$. We may expect that α may be large if β

is large since

$$\dot{x} = |t|^{-\alpha} \sin |t|^{-\beta}$$

is a particular case of (18.7) and

$$\lim_{t \to 0+} \int_t^1 \tau^{-\alpha} \sin \tau^{-\beta}\, d\tau$$

exists if $\alpha < 1 + \beta$.

18.3. Lemma. *Let* $\xi \in \mathbb{R}$, $\xi > -1$, $0 \leq S < T \leq 1$. *Then*

$$\int_S^T \sigma^\xi\, d\sigma \leq \frac{2+\xi}{1+\xi}\,(T-S)^{\min\{1,1+\xi\}}\,. \qquad (18.8)$$

In particular,

$$\int_S^T \sigma^{\beta-\alpha}\, d\sigma \leq \frac{2+\beta-\alpha}{1+\beta-\alpha}\,(T-S)^\eta\,, \qquad (18.9)$$

$$\int_S^T \sigma^{1+\beta-2\alpha}\, d\sigma \leq \frac{3+\beta-2\alpha}{2+\beta-2\alpha}\,(T-S)^\eta\,. \qquad (18.10)$$

Proof. If $\xi \geq 0$ then

$$\int_S^T \sigma^\xi\, d\sigma \leq (T-S)\,.$$

If $-1 < \xi < 0$ then

$$\int_S^T \sigma^\xi\, d\sigma \leq \int_0^{T-S} \sigma^\xi\, d\sigma \leq \frac{1}{1+\xi}(T-S)^{1+\xi}$$

and (18.8) holds. (18.9) and (18.10) follow from (18.8).   □

18.4. Lemma. *Let* $0 < a < b \leq 1$ *and let* $u: [a,b] \to B(\widehat{R})$ *be a (classical) solution of*

$$\dot{x} = t^{-\alpha} h(x, t^{-\beta})\,. \qquad (18.11)$$

Then

$$u(T) - u(S) = \int_S^T \sigma^{-\alpha} h(u(\sigma), \sigma^{-\beta})\, d\sigma \qquad (18.12)$$

for $a \leq S < T \leq b$ and

$$
\left.
\begin{aligned}
u(T) &- u(S) \\
&= \frac{1}{\beta} S^{1+\beta-\alpha} H(u(S), S^{-\beta}) \\
&\quad - \frac{1}{\beta} T^{1+\beta-\alpha} H(u(T), T^{-\beta}) \\
&\quad + \frac{1+\beta-\alpha}{\beta} \int_S^T \sigma^{\beta-\alpha} H(u(\sigma), \sigma^{-\beta}) \, d\sigma \\
&\quad + \frac{1}{\beta} \int_S^T \sigma^{1+\beta-2\alpha} D_1 H(u(\sigma), \sigma^{-\beta}) \, h(u(\sigma), \sigma^{-\beta}) \, d\sigma
\end{aligned}
\right\} \qquad (18.13)
$$

for $a \leq S < T \leq b$, $D_1 H$ being the differential of $H(x, t)$ with respect to x.

Proof. (18.12) is an immediate consequence of (18.11). Further, (18.13) holds since

$$
\frac{d}{dt}(t^{1+\beta-\alpha} H(u(t), t^{-\beta})) = (1+\beta-\alpha) t^{\beta-\alpha} H(u(t), t^{-\beta})
$$

$$
+ t^{1+\beta-\alpha} D_1 H(u(t), t^{-\beta}) \, \dot{u}(t) - \beta t^{-\alpha} h(u(t), t^{-\beta})
$$

and

$$
\dot{u}(t) = t^{-\alpha} h(u(t), t^{-\beta}). \qquad \square
$$

18.5. Lemma. *Let $0 < a < b \leq 1$ and let $u : [a, b] \to B(\widehat{R})$ be a solution of (18.11). Then*

$$
\|u(T) - u(S)\| \leq \frac{2\nu}{\beta} T^\eta + \varkappa_1 (T-S)^\eta \quad \text{for} \quad a \leq S < T \leq b \qquad (18.14)
$$

where

$$
\varkappa_1 = \frac{\nu}{\beta}(2+\beta-\alpha) + 2 \frac{\nu^2}{\beta} \frac{3+\beta-2\alpha}{2+\beta-2\alpha}. \qquad (18.15)
$$

Moreover,

$$
\|u(S)\| \leq \|u(T)\| + \frac{2\nu}{\beta} T^\eta + \varkappa_1 T^\eta. \qquad (18.16)
$$

Proof.　Let $a \leq S < T \leq b$. By (18.13), (18.9), (18.10)

$$\|u(T) - u(S)\| \leq \frac{2\nu}{\beta} T^{1+\beta-\alpha} + \frac{\nu}{\beta}(1+\beta-\alpha)\int_S^T \sigma^{\beta-\alpha}\,d\sigma$$

$$+ \frac{\nu^2}{\beta}\int_S^T \sigma^{1+\beta-2\alpha}\,d\sigma$$

$$\leq \frac{2\nu}{\beta} T^{1+\beta-\alpha} + \varkappa_1 (T-S)^\eta$$

and (18.14) holds since $0 < \eta \leq 2+\beta-2\alpha < 1+\beta-\alpha$. (18.16) is a consequence of (18.14). □

18.6. Lemma. *Let* $0 < c \leq d \leq 1$, $y \in X$,

$$\|y\| + \frac{2\nu}{\beta} d^\eta + \varkappa_1 d^\eta \leq \widehat{R}. \tag{18.17}$$

Then there exists an SKH-solution u *of* (18.7) *on* $[0,d]$ *such that* $u(c) = y$.
Moreover, $\lim_{t\to 0+} u(t)$ *exists.*

Proof.　By classical results on ordinary differential equations there exist
$P, Q, 0 < P < c < Q \leq d$, $u : [P, Q] \to D(\widehat{R})$ such that u is a classical solution
of (18.7), $u(c) = y$. By Lemma 18.5

$$\|u(t) - y\| \leq \left(\frac{2\nu}{\beta} + \varkappa_1\right) t^\eta \leq \left(\frac{2\nu}{\beta} + \varkappa_1\right) d^\eta \quad \text{for } c \leq t \leq Q$$

and

$$\|u(t) - y\| \leq \left(\frac{2\nu}{\beta} + \varkappa_1\right) c^\eta \quad \text{for } P \leq t \leq c.$$

By standard arguments and by (18.17) and Lemma 18.5 there exists a
classical solution u of (18.7) such that $u(t)$ is defined for $0 < t \leq d$, $u(c) = y$
and

$$\left.\begin{array}{l}\|u(t)\| \leq \|y\| + \left(\dfrac{2\nu}{\beta} + \varkappa_1\right) d^\eta \quad \text{for } 0 < t \leq d, \\[2mm] \|u(t) - u(S)\| \leq \left(\dfrac{2\nu}{\beta} + \varkappa_1\right) S^\eta \quad \text{for } 0 < t \leq S \leq d.\end{array}\right\} \tag{18.18}$$

u is an SKH-solution of (18.7) on $[a,d]$ for $0 < a < d$. (18.18) implies that
$\lim_{t\to 0+} u(t)$ exists. Define $u(0) = \lim_{t\to 0+} u(t)$. Then u is an SKH-solution
of (18.7) on $[0,d]$ by Theorem 15.8. □

18.7. Lemma. *Let*

$$0 < d \le \tfrac{1}{2}, \quad \frac{\nu}{\beta} d^{1+\beta-\alpha} \le \tfrac{1}{2} \tag{18.19}$$

and let $u : [0,d] \to B(\widehat{R})$, $\overline{u} : [0,d] \to B(\widehat{R})$ *be SKH-solutions of* (18.7). *Then*

$$\left.\begin{array}{l} \|u(T) - \overline{u}(T)\| \le 3\, \|u(S) - \overline{u}(S)\|\ \exp(2\,\varkappa_1\,(T-S)^{\eta}) \\[2mm] \qquad\qquad\qquad\qquad\qquad \text{for}\ \ 0 \le S < T \le d. \end{array}\right\} \tag{18.20}$$

Proof. *u* fulfils (cf. (18.13))

$$u(T) + \frac{1}{\beta} T^{1+\beta-\alpha} H(u(T), T^{-\beta})$$

$$= u(S) + \frac{1}{\beta} S^{1+\beta-\alpha} H(u(S), S^{-\beta})$$

$$+ \frac{1+\beta-\alpha}{\beta} \int_S^T \sigma^{\beta-\alpha}\, H(u(\sigma), \sigma^{-\beta})\, \mathrm{d}\sigma$$

$$+ \frac{1}{\beta} \int_S^T \sigma^{1+\beta-2\alpha}\, D_1 H(u(\sigma), \sigma^{-\beta})\, h(u(\sigma), \sigma^{-\beta})\, \mathrm{d}\sigma.$$

Subtracting the same equation with u replaced by $\overline{u}$ we obtain

$$u(T) - \overline{u}(T) + \frac{1}{\beta} T^{1+\beta-\alpha} [H(u(T), T^{-\beta}) - H(\overline{u}(T), T^{-\beta})]$$

$$= u(S) - \overline{u}(S) + \frac{1}{\beta} S^{1+\beta-\alpha} [H(u(S), S^{-\beta}) - H(\overline{u}(S), S^{-\beta})]$$

$$+ \frac{1+\beta-\alpha}{\beta} \int_S^T \sigma^{\beta-\alpha} [H(u(\sigma), \sigma^{-\beta}) - H(\overline{u}(\sigma), \sigma^{-\beta})]\, \mathrm{d}\sigma$$

$$+ \frac{1}{\beta} \int_S^T \sigma^{1+\beta-2\alpha} [D_1 H(u(\sigma), \sigma^{-\beta})\, h(u(\sigma), \sigma^{-\beta})$$

$$- D_1 H(\overline{u}(\sigma), \sigma^{-\beta}) h(\overline{u}(\sigma, \sigma^{-\beta})]\, \mathrm{d}\sigma.$$

Hence (cf. (18.5), (18.19))

$$\tfrac{1}{2} \|u(T) - \overline{u}(T)\| \le \tfrac{3}{2} \|u(S) - \overline{u}(S)\|$$

$$+ \frac{\nu}{\beta} (1+\beta-\alpha) \int_S^T \sigma^{\beta-\alpha} \|u(\sigma) - \overline{u}(\sigma)\|\, \mathrm{d}\sigma$$

$$+ \frac{2\nu^2}{\beta} \int_S^T \sigma^{1+\beta-2\alpha} \|u(\sigma) - \overline{u}(\sigma)\|\, \mathrm{d}\sigma.$$

Lemma A.1 implies (cf. also (18.9), (18.10), (18.15)) that (18.20) holds. □

18.8. Lemma. *Let $y \in X$, $\|y\| < \widehat{R}$ and let (18.16) hold. Then there exists an SKH-solution $u_1 : [0, d] \to B(\widehat{R})$ of (18.7) on $[0, d]$ fulfilling $u_1(0) = y$.*

Proof.   By Lemma 18.6 there exists an SKH-solution v_i of (18.7) on $[0, d]$, $v_i(d \, 2^{-i}) = y$ for $i \in \mathbb{N}$. By Lemma 18.5

$$\|v_{i+1}(d \, 2^{-i}) - v_{i+1}(d \, 2^{-i-i})\| \leq (\frac{2\,\nu}{\beta} + \varkappa_1) \, (d \, 2^{-i})^{\eta}.$$

Hence

$$\|v_{i+1}(d \, 2^{-i}) - v_i(d \, 2^{-i})\| \leq \left(\frac{2\,\nu}{\beta} + \varkappa_1\right) (d \, 2^{-i})^{\eta}$$

since $v_{i+1}(d \, 2^{-i-1}) = y = v_i(d \, 2^{-i})$. By Lemma 18.7

$$\left.\begin{array}{r}\|v_{i+1}(t) - v_i(t)\| \leq 3 \left(\dfrac{2\,\nu}{\beta} + \varkappa_1\right) (d \, 2^{-i})^{\eta} \exp(2 \, \varkappa_1 \, d^{\eta}) \\[2mm] \text{for } d \, 2^{-i} \leq t \leq d, \ i \in \mathbb{N}.\end{array}\right\} \quad (18.21)$$

By Lemma 18.5

$$\|v_{i+1}(t) - v_{i+1}(d \, 2^{-i})\| \leq \left(\frac{2\,\nu}{\beta} + \varkappa_1\right) (d \, 2^{-i})^{\eta},$$

$$\|v_i(t) - v_i(d \, 2^{-i})\| \leq \left(\frac{2\,\nu}{\beta} + \varkappa_1\right) (d \, 2^{-i})^{\eta},$$

$$\|v_{i+1}(t) - v_i(t)\|$$

$$\leq 2 \left(\frac{2\,\nu}{\beta} + \varkappa_1\right) (d \, 2^{-i})^{\eta} + \|v_{i+1}(d \, 2^{-i}) - v_i(d \, 2^{-i})\|$$

$$\leq 3 \left(\frac{2\,\nu}{\beta} + \varkappa_1\right) (d \, 2^{-i})^{\eta}$$

for $0 < t \leq d \, 2^{-i}$.

The above inequalities hold for $t = 0$ as well since $\lim\limits_{t \to 0+} v_i(t)$ exists for $i \in \mathbb{N}$. Hence

$$\|v_{i+1}(t) - v_i(t)\| \leq 3 \, \varkappa_1 (d \, 2^{-i})^{\eta} \text{ for } 0 \leq t \leq d \, 2^{-i}, \, i \in \mathbb{N}. \qquad (18.22)$$

(18.21) together with (18.22) implies that the sequence v_i, $i \in \mathbb{N}$, is convergent uniformly on $[0, d]$. Put $u_1(t) = \lim\limits_{i \to \infty} v_i(t)$ for $t \in [0, d]$. Obviously, u_1 is continuous and $u_1(0) = y$.

It remains to prove that u_1 is an SKH-solution of (18.7). It is sufficient to prove that

$$u_1(T) - u_1(S) = (\text{SKH}) \int_S^T g(u_1(t), t)\, dt \qquad (18.23)$$

for $0 \le S \le T \le d$ (cf. (18.6) and Theorem 17.3 (i), (ii)). Equality (18.23) holds for $0 < S \le T \le d$ since $g(u_1(t), t)$ is continuous for $0 < t \le d$. The integral $(\text{SKH}) \int_0^d g(u_1(t), t)\, dt$ exists by Theorem 14.20 where $U(\tau, t) = g(u_1(\tau), \tau)\, t$ for $\tau \in [0, d]$, $t \in \mathbb{R}$, and $w = u_1(d) - u_1(0)$ since $U(u_1(0), t) = 0$ and

$$\lim_{S \to 0+} \left(-(\text{SKH}) \int_S^d g(u_1(t), t)\, dt + u_1(d) - u_1(0) \right)$$
$$= \lim_{S \to 0+} \left(-u_1(d) + u_1(S) + u_1(d) - u_1(0) \right) = \lim_{S \to 0+} (u_1(S) - u_1(0)) = 0.$$

(18.23) holds for $0 \le S \le T \le d$ and Lemma 18.8 is correct. $\qquad \square$

18.9. Theorem. *Let* $y \in X$, $\|y\| < \widehat{R}$ *and let* (18.19) *hold. Then there exists an SKH-solution* $u : [-d, d] \to B(\widehat{R})$ *of* (18.7) *on* $[-d, d]$ *fulfilling* $u(0) = y$. *Moreover, u is unique.*

Proof. By Lemma 18.8 there exists an SKH-solution $u_1 : [0, d] \to B(\widehat{R})$ of (18.7) such that $u_1(0) = y$.

Let
$$f(x, \tau) = -g(x, -\tau) \quad \text{for } x \in B(\widehat{R}), \ \tau \in [0, 1].$$
Lemma 18.8 may be applied to the equation

$$\dot{x} = f(x, t) \qquad (18.24)$$

since (cf. (18.6)) $f(x, t) = -(-t)^{-\alpha} h(x, (-t)^{-\beta})$ and h fulfils (18.2)–(18.5). Hence there exists an SKH-solution w of (18.23) on $[0, d]$ such that $w(0) = y$. Put $u_2(\tau) = w(-\tau)$ for $\tau \in [-d, 0]$. By Lemma 17.7 (where $a = 0$, $b = d$) u_2 is an SKH-solution of (18.7) on $[-d, 0]$, $u_2(0) = y$. Let

$$u(t) = \begin{cases} u_1(t) & \text{if } t \in [0, d], \\ u_2(t) & \text{if } t \in [-d, 0]. \end{cases}$$

u is an SKH solution of (18.7) on $[-d, d]$ by Lemma 15.6. The uniqueness of u_1 is a consequence of Lemma 18.7. Similarly, also u_2 is unique. Hence u is unique. $\qquad \square$

Chapter 19

Integration and Strong Integration

19.1. Notation. Let $\operatorname{Dom} U \subset \mathbb{R}^2$, $U : \operatorname{Dom} U \to X$.

19.2. Definition. U is KH-*integrable* on $[a, b]$ if there exist $\delta_0 : [a, b] \to \mathbb{R}^+$ and $\Gamma(a, b) \in X$ such that

$$\{(\tau, t) \in [a, b]^2 ; \tau - \delta_0(\tau) \le t \le \tau + \delta_0(\tau)\} \subset \operatorname{Dom} U \qquad (19.1)$$

and for every $\varepsilon > 0$ there exists $\delta : [a, b] \to \mathbb{R}^+$ such that

$$\left\| \Gamma(a, b) - \sum_{\mathcal{A}} (U(\tau, t) - U(\tau, \bar{t})) \right\| \le \varepsilon \qquad (19.2)$$

for every δ-fine partition $\mathcal{A} = \{([\bar{t}, t], \tau)\}$ of $[a, b]$.

19.3. Remark. $\Gamma(a, b)$ is unique. Indeed, if there exists $\widehat{\Gamma}(a, b) \neq \Gamma(a, b)$ and for every $\varepsilon > 0$ there exists $\widehat{\delta} : [a, b] \to \mathbb{R}^+$ such that $\widehat{\delta}(\tau) \le \delta(\tau)$ for $\tau \in [a, b]$ and

$$\left\| \widehat{\Gamma}(a, b) - \sum_{\mathcal{A}} (U(\tau, t) - U(\tau, \bar{t})) \right\| \le \varepsilon$$

for every $\widehat{\delta}$-fine partition $\mathcal{A}$ of $[a, b]$, then a contradiction is obtained e.g. for $\varepsilon = \frac{1}{3} \|\Gamma(a, b) - \widehat{\Gamma}(a, b)\|$). $\Gamma(a, b)$ is denoted by

$$(\mathrm{KH}) \int_a^b \mathrm{D}_t U(\tau, t)$$

and called the *Kurzweil–Henstock integral* (KH-*integral*) of U over $[a, b]$.

19.4. Theorem. *Let U be SKH-integrable on $[a, b]$, $u : [a, b] \to X$ being its primitive. Then U is KH-integrable on $[a, b]$ and*

$$(\mathrm{KH}) \int_S^T \mathrm{D}_t U(\tau, t) = (\mathrm{SKH}) \int_S^T \mathrm{D}_t U(\tau, t) .$$

Proof. Let U be SKH-integrable on $[a, b]$ and let u be its primitive. Then for $\varepsilon > 0$ there exists $\delta : [a, b] \to \mathbb{R}^+$ such that

$$\sum_{\mathcal{A}} \|u(t) - u(\bar{t}) - U(\tau, t) + U(\tau, \bar{t})\| \leq \varepsilon$$

for any δ-fine partition $\mathcal{A} = \{([\bar{t}, t], \tau)\}$ of $[a, b]$. Hence

$$\|u(b) - u(a) - \sum_{\mathcal{A}} (U(\tau, t) - U(\tau, \bar{t}))\| \leq \varepsilon$$

and the theorem is valid.   □

19.5. Remark. Theorem 19.4 can be converted if $\dim X < \infty$ (cf. Corollary 19.14).

19.6. Lemma. *Let U be KH-integrable on $[a, b]$. Then*

$$U \quad \text{is KH-integrable on } [S, T] \quad \text{for } [S, T] \subset [a, b]. \tag{19.3}$$

Moreover,

$$\text{(KH)} \int_S^T \mathrm{D}_t U(\tau, t) = \text{(KH)} \int_S^{\widehat{T}} \mathrm{D}_t U(\tau, t) + \text{(KH)} \int_{\widehat{T}}^T \mathrm{D}_t U(\tau, t) \tag{19.4}$$

for $a \leq S \leq \widehat{T} \leq T \leq b$.

Proof. Let $\varepsilon > 0$ and let $\delta : [a, b] \to \mathbb{R}^+$ correspond to ε by Definition 19.2, $[S, T] \subset [a, b]$. There exist

a δ-fine partition $\mathcal{A} = \{([\bar{t}, t], \tau)\}$ of $[a, S]$

and

a δ-fine partition $\mathcal{C} = \{([\bar{t}, t], \tau)\}$ of $[T, b]$.

If $\mathcal{B}_1 = \{([\bar{t}, t], \tau)\}$ and $\mathcal{B}_2 = \{([\bar{t}, t], \tau)]$ are δ-fine partitions of $[S, T]$, then $\mathcal{A} \cup \mathcal{B}_1 \cup \mathcal{C}$ and $\mathcal{A} \cup \mathcal{B}_2 \cup \mathcal{C}$ are δ-fine partitions of $[a, b]$. Hence

$$\left\| \Gamma(a, b) - \sum_{\mathcal{A} \cup \mathcal{B}_1 \cup \mathcal{C}} (U(\tau, t) - U(\tau, \bar{t})) \right\| \leq \varepsilon,$$

$$\left\| \Gamma(a, b) - \sum_{\mathcal{A} \cup \mathcal{B}_2 \cup \mathcal{C}} (U(\tau, t) - U(\tau, \bar{t})) \right\| \leq \varepsilon,$$

and

$$\left\| \sum_{\mathcal{B}_1}(U(\tau,t)-U(\tau,\bar{t})) - \sum_{\mathcal{B}_2}(U(\tau,t)-U(\tau,\bar{t})) \right\| \le 2\varepsilon. \qquad (19.5)$$

(19.5) implies that there exists $\Gamma(S,T)$ such that

$$\left\| \Gamma(S,T) - \sum_{\mathcal{B}}(U(\tau,t)-U(\tau,\bar{t})) \right\| \le 2\varepsilon$$

for every δ-fine partition $\mathcal{B} = \{([\bar{t},t],\tau)\}$ of $[S,T]$. (19.3) is correct.

Let $a \le S \le \widehat{T} \le T \le b$. By (19.3) U is KH-integrable on $[S,\widehat{T}]$, on $[\widehat{T},T]$ and on $[S,T]$. Therefore there exists $\delta:[S,T]\to\mathbb{R}^+$ such that

$$\left. \begin{array}{l} \left\| \Gamma(S,\widehat{T}) - \displaystyle\sum_{\mathcal{B}_1}(U(\tau,t)-U(\tau,\bar{t})) \right\| \le \varepsilon \\[2mm] \quad \text{for every } \delta-\text{fine partition } \mathcal{B}_1 = \{([\widehat{t},t],\tau)\} \text{ of } [S,\widehat{T}], \end{array} \right\} \quad (19.6)$$

$$\left. \begin{array}{l} \left\| \Gamma(\widehat{T},T) - \displaystyle\sum_{\mathcal{B}_2}(U(\tau,t)-U(\tau,\bar{t})) \right\| \le \varepsilon \\[2mm] \quad \text{for every } \delta-\text{fine partition } \mathcal{B}_2 = \{([\widehat{t},t],\tau)\} \text{ of } [\widehat{T},T], \end{array} \right\} \quad (19.7)$$

and

$$\left. \begin{array}{l} \left\| \Gamma(S,T) - \displaystyle\sum_{\mathcal{B}}(U(\tau,t)-U(\tau,\bar{t})) \right\| \le \varepsilon \\[2mm] \quad \text{for every } \delta\text{-fine partition } \mathcal{B} = \{([\widehat{t},t],\tau)\} \text{ of } [S,T]. \end{array} \right\} \quad (19.8)$$

Let $\mathcal{B}_1$ be a δ-fine partition of $[S,\widehat{T}]$ and let $\mathcal{B}_2$ be a δ-fine partition of $[\widehat{T},T]$. Then $\mathcal{B} = \mathcal{B}_1 \cup \mathcal{B}_2$ is a δ-fine partition of $[S,T]$ and (19.6)–(19.8) imply that

$$\|\Gamma(S,\widehat{T})+\Gamma(\widehat{T},T)-\Gamma(S,T)\| \le 3\varepsilon. \qquad (19.9)$$

(19.4) holds by (19.9) since $\varepsilon > 0$ is arbitrary. $\qquad\square$

19.7. Remark. Let U be KH-integrable on $[a,b]$. Then $u:[a,b]\to X$ is called a KH-*primitive of U on $[a,b]$* or simply a *primitive of U* if

$$u(T) - u(S) = (\text{KH})\int_S^T D_t U(\tau,t) \quad \text{for } [S,T] \subset [a,b].$$

By Lemma 19.6 $(\text{KH})\int_a^T D_t U(\tau,t)$ exists for $T \in [a,b]$. Put

$$v(T) = (\text{KH})\int_a^T D_t U(\tau,t) \quad \text{for } T \in [a,b].$$

Then v is a primitive of U. On the other hand, if u is a primitive of U then $u(T) - u(a) = v(T)$ for $T \in [a, b]$.

19.8. Remark. Similarly to Remark 14.16 we put

$$(KH) \int_T^S D_t U(\tau, t) = -(KH) \int_S^T D_t U(\tau, t) \quad \text{for } a \le S \le T \le b.$$

19.9. Lemma. *Let U be KH-integrable on $[a, b]$, $\varepsilon > 0$ and let $\delta : [a, b] \to \mathbb{R}^+$ correspond to ε by Definition 19.2. Then*

$$\left\| \sum_{\mathcal{D}} (\Gamma(\bar{s}, s) - U(\sigma, s) + U(\sigma, \bar{s})) \right\| \le 2\varepsilon \tag{19.10}$$

for every δ-fine system $\mathcal{D} = \{([\bar{s}, s], \sigma)\}$ in $[a, b]$.

Proof. Let $D = \{([\bar{s}_i, s_i], \sigma_i); i = 1, 2, \dots, k\}$ be a δ-fine system in $[a, b]$. For simplicity assume that

$$a = \bar{s}_1 < s_1 < \bar{s}_2 < s_2 < \cdots < \bar{s}_k < s_k = b.$$

By Lemma 19.6, U is KH-integrable on $[s_{i-1}, \bar{s}_i]$ for $i = 2, 3, \dots, k$. Hence there exist $\delta_i : [s_{i-1}, \bar{s}_i] \to \mathbb{R}^+$ such that $\delta_i(\tau) \le \delta(\tau)$ for $\tau \in [s_{i-1}, \bar{s}_i]$ and

$$\left\| \Gamma(s_{i-1}, \bar{s}_i) - \sum_{\mathcal{E}_i} (U(\rho, r) - U(\rho, \bar{r})) \right\| \le \frac{\varepsilon}{k}$$

for every δ_i-fine partition $\mathcal{E}_i = \{([\bar{r}, r], \rho)\}$ of $[s_{i-1}, \bar{s}_i]$. By Lemma 14.4 there exists a δ_i-fine partition $\widehat{\mathcal{E}}_i = \{([\bar{r}, r], \rho)\}$ of $[s_{i-1}, \bar{s}_i]$. Then $\mathcal{D} \cup (\bigcup_{i=2}^{k} \widehat{\mathcal{E}}_i)$ is a δ-fine partition of $[a, b]$,

$$\left\| \sum_{i=1}^{k} (\Gamma(\bar{s}_i, s_i) - U(\sigma_i, \bar{s}_i) + U(\sigma_i, s_i)) \right.$$
$$\left. + \sum_{i=2}^{k} \left(\Gamma(s_{i-1}, \bar{s}_i) - \sum_{\widehat{\mathcal{E}}_i} (U(\rho, r) - U(\rho, \bar{r})) \right) \right\| \le \varepsilon,$$

$$\left\| \Gamma_i(s_{i-1}, \bar{s}_i) - \sum_{\widehat{\mathcal{E}}_i} (U(\rho, r) - U(\rho, \bar{r})) \right\| \le \frac{\varepsilon}{k}$$

and (19.10) holds.   ⊓

19.10. Lemma. *Let $\varphi\colon [a,b] \to \mathbb{R}$ be nondecreasing, let $\delta_0\colon [a,b] \to \mathbb{R}^+$ and let U be KH-integrable on $[a,b]$. Assume that*

$$\|U(\tau,t) - U(\tau,\bar{t})\| \le \varphi(t) - \varphi(\bar{t}) \quad \text{for } \tau,t,\bar{t} \in [a,b],\ \bar{t} \le \tau \le t\,,$$

and

$$\tau - \delta_0(\tau) \le \bar{t} \le \tau \le t \le \tau + \delta_0(\tau)\,.$$

Then

$$\left\| (\text{KH}) \int_S^T D_t U(\tau,t) \right\| \le \varphi(T) - \varphi(S) \quad \text{for } a \le S < T \le b. \tag{19.11}$$

Proof. Let $\varepsilon > 0$, $[S,T] \subset [a,b]$. By Definition 19.2 and Lemmas 19.6, 14.4 there exists a δ_0-fine partition $\mathcal{T} = \{([\bar{t},t],\tau)\}$ of $[S,T]$ such that

$$\left\| (\text{KH}) \int_S^T D_t U(\tau,t) - \sum_{\mathcal{T}} (U(\tau,t) - U(\tau,\bar{t})) \right\| \le \varepsilon\,.$$

Hence

$$\left\| (\text{KH}) \int_S^T D_t U(\tau,t) \right\| \le \varphi(T) - \varphi(S) + \varepsilon$$

and (19.11) holds, since ε is arbitrary positive. $\qquad\square$

19.11. Remark. Let $U\colon [a,b]^2 \to X$. Put

$$-U^*(-\tau,-t) = U(\tau,t) \quad \text{for } (\tau,t) \in [a,b]^2\,.$$

Then $U^*\colon [-b,-a]^2 \to X$.

Let $\mathcal{T} = \{([t_{i-1},t_i],\tau_i),\ i=1,2,\ldots,k\}$ be a partition of $[a,b]$. Then

$$\left.\begin{aligned}
&\sum_{i=1}^k (U(\tau_i,t_i) - U(\tau_i,t_{i-1}))\\
&\quad = -\sum_{i=1}^k (U^*(-\tau_i,-t_i) - U^*(-\tau_i,-t_{i-1}))\,.
\end{aligned}\right\} \tag{19.12}$$

Put $\mathcal{T}^* = \{([-t_i,-t_{i-1}],\tau_i);\ i=1,2,\ldots,k\}$. Then $\mathcal{T}^*$ is a partition of $[-b,-a]$. By (19.11)

$$\sum_{i=1}^k (U(\tau_i,t_i) - U(\tau_i,t_{i-1})) = \sum_{i=1}^k (U^*(-\tau_i,-t_{i-1}) - U^*(-\tau_i,-t_i))\,.$$

Let $\delta:[a,b]\to\mathbb{R}^+$, $\delta^*:[-b,-a]\to\mathbb{R}^+$, $\delta^*(-\tau)=\delta(\tau)$ for $\tau\in[a,b]$. If $\mathcal{T}$ is δ-fine, then $\mathcal{T}^*$ is δ^*-fine and vice versa. We may conclude that

$$(\mathrm{KH})\int_{-b}^{-a} D_t U^*(\tau,t) = (\mathrm{KH})\int_a^b D_t U(\tau,t) \quad \text{for } a<b \tag{19.13}$$

if one of the integrals exists.

If $U(\tau,t)=\xi(t)\,\Phi(t)$, put $\xi^*(-\tau)=\xi(\tau)$ and $\Phi^*(-t)=-\,\Phi(t)$. Then $-U^*(-\tau,-t)=U(\tau,t)$ and

$$(\mathrm{KH})\int_{-b}^{-a} D_t\xi^*(\tau)\,\Phi^*(t) = (\mathrm{KH})\int_a^b D_t\xi(\tau)\,\Phi(t) \tag{19.14}$$

if one of the integrals exists, and

$$(\mathrm{SKH})\int_{-b}^{-a} D_t\xi^*(\tau)\,\Phi^*(t) = (\mathrm{SKH})\int_a^b D_t\xi(t)\,\Phi(t) \tag{19.15}$$

if one of the integrals exists.

19.12. Lemma. *Let $n\in\mathbb{N}$, $X=\mathbb{R}^n$,*

$$\|x\| = \sum_{i=1}^n |x_i| \quad \text{for } x = (x_1,x_2,\ldots,x_n)\in\mathbb{R}^n \tag{19.16}$$

and

$$\mathcal{L} = \{-1,1\}^n. \tag{19.17}$$

For $L\in\mathcal{L}$ (i.e. $L=(L_1,L_2,\ldots,L_n)$, $|L_i|=1$) put

$$Q_L = \{x\in\mathbb{R}^n; L_i x_i \geq 0 \quad \text{for } i=1,2,\ldots,n\}. \tag{19.18}$$

Then

$$\|z\| = \sum_{i=1}^n L_i z_i \qquad\qquad \text{for } z = (z_1,z_2,\ldots,z_n)\in Q_L, \tag{19.19}$$

$$\left\|\sum_{z\in Z} z\right\| = \sum_{z\in Z} \|z\| \qquad\qquad \text{if } Z\subset Q_L \text{ is finite}, \tag{19.20}$$

$$\sum_{z\in Z} \|z\| \leq 2^n \max_{L\in\mathcal{L}} \left\|\sum_{z\in Z\cap Q_L} z\right\| \quad \text{if } Z\subset Q_L \text{ is finite}. \tag{19.21}$$

Proof. (19.19) follows by (19.16)–(19.18). Further, (19.20) is a consequence of (19.19). Finally, let $Z \subset \mathbb{R}^n$ be finite. Then

$$\sum_{z \in Z} \|z\| \leq \sum_{L \in \mathcal{L}} \sum_{z \in Z \cap Q_L} \|z\| \leq \sum_{L \in \mathcal{L}} \left\| \sum_{z \in Z \cap Q_L} z \right\|$$

$$\leq 2^n \max_{L \in \mathcal{L}} \left\| \sum_{z \in Z \cap Q_L} z \right\|.$$

Hence (19.21) holds. □

19.13. Lemma. *Assume that $n \in \mathbb{N}$ and that $U : \mathrm{Dom}\, U \to \mathbb{R}^n$ is KH-integrable on $[a, b]$. Given $\varepsilon > 0$, let $\delta : [a, b] \to \mathbb{R}^+$ be such that*

$$\left\| \Gamma(a, b) - \sum_{\mathcal{A}} (U(\tau, t) - U(\tau, \bar{t})) \right\| \leq \varepsilon$$

for every δ-fine partition $\mathcal{A} = \{([\bar{t}, t], \tau)\}$ of $[a, b]$.

Then

$$\sum_{\mathcal{A}} \| \Gamma(\bar{t}, t) - (U(\tau, t) - U(\tau, \bar{t})) \| \leq 2^n \, \varepsilon \qquad (19.22)$$

for every δ-fine partition $\mathcal{A} = \{([\bar{t}, t], \tau)\}$ of $[a, b]$.

Proof. Let $\varepsilon > 0$, let δ correspond to ε by Definition 19.2, let the partition $\mathcal{A} = \{([\bar{s}, s], \sigma)\}$ of $[a, b]$ be δ-fine and let

$$Z = \{z = \Gamma(\bar{s}, s) - U(\sigma, s) + U(\sigma, \bar{s}) \, ; ([\bar{s}, s], \sigma) \in \mathcal{A}\}.$$

Then

$$\left\| \sum_{z \in Z \cap Q_L} z \right\| \leq \varepsilon$$

and (19.22) is true by (19.21). □

19.14. Corollary. *Let $U : \mathrm{Dom}\, U \to \mathbb{R}^n$ be KH-integrable on $[a, b]$. Then U is SKH-integrable on $[a, b]$ and*

$$(\mathrm{SKH}) \int_a^b \mathrm{D}_t U(\tau, t) = (\mathrm{KH}) \int_a^b \mathrm{D}_t U(\tau, t).$$

Chapter 20

A class of Strong Kurzweil Henstock-integrable functions

The main result of Chapter 20 implies that the product $\psi(\tau)\,\varphi(t)$ is SKH-integrable if ψ is a regulated function and φ is a function of bounded variation.

20.1. Notation. Let $[c,d] \subset [a,b] \subset \mathbb{R}$, $U:[a,b]^2 \to X$, $\Psi:[a,b] \to \mathbb{R}$, $\Phi:[a,b] \to \mathbb{R}$, $\varepsilon > 0$. Assume that

$$\Psi \quad \text{is a regulated function}, \tag{20.1}$$

$$\Phi \quad \text{is nondecreasing}, \ \Phi(b) - \Phi(a) > 0, \tag{20.2}$$

$$\|U(\tau,t) - U(\tau,s)\| \le |\Phi(t) - \Phi(s)| \quad \text{for} \ t,s,\tau \in [a,b], \tag{20.3}$$

$$\left. \begin{aligned} &\|U(\tau,t) - U(\tau,s) - U(\sigma,t) + U(\sigma,s)\| \\ &\quad \le (\Psi(\tau) - \Psi(\sigma))\,(\Phi(t) - \Phi(s)) \quad \text{for} \ t,s,\tau,\sigma \in [a,b]. \end{aligned} \right\} \tag{20.4}$$

Define sets $\mathbf{A}(\varepsilon) = \mathbf{A}(\varepsilon)(\varepsilon,c,d)$, $\mathbf{B}(\varepsilon) = \mathbf{B}(\varepsilon)(\varepsilon,c,d)$ by

$$\left. \begin{aligned} \mathbf{A}(\varepsilon) = \ &\{\tau; \ c \le \tau < d, \ |\Psi(\tau+) - \Psi(\tau)| \ge \varepsilon\} \\ &\cup \{\tau; \ c < \tau \le d, \ |\Psi(\tau) - \Psi(\tau-)| \ge \varepsilon\}, \end{aligned} \right\} \tag{20.5}$$

$$\left. \begin{aligned} \mathbf{B}(\varepsilon) = \ &\{\tau; \ c \le \tau < d, \ |\Psi(\tau+) - \Psi(\tau)| < \varepsilon\} \\ &\cap \{\tau; \ c < \tau \le d, \ |\Psi(\tau) - \Psi(\tau-)| < \varepsilon\}. \end{aligned} \right\} \tag{20.6}$$

Obviously,

$$\mathbf{B}(\varepsilon) = [c,d] \setminus \mathbf{A}(\varepsilon). \tag{20.7}$$

and (20.1) implies that $\mathbf{A}(\varepsilon)$ is a finite set. Denote by $m(\varepsilon)$ the number of elements of $\mathbf{A}(\varepsilon)$.

20.2. Definition. A partition $\mathcal{T} = \{([\bar{t}, t], \tau)\}$ of $[c, d]$ is called *simple* if $\tau = \bar{t}$ or $\tau = t$ for $([\bar{t}, t], \tau) \in \mathcal{T}$ and if $c < \bar{t} < t < d$ whenever $c < \tau < d$.

20.3. Remark. Let $\delta : [c, d] \to \mathbb{R}^+$. Let $\mathcal{T} = \{([\bar{t}, t], \tau)\}$ be a partition of $[c, d]$ such that $c < \bar{t} < t < d$ whenever $c < \tau < d$. Modify $\mathcal{T}$ in the following way: keep elements $([\bar{t}, t], \tau)$ such that $\tau = \bar{t}$ or $\tau = t$ and replace every element $([\bar{t}, t], \tau)$ such that $\bar{t} < \tau < t$ by two elements $([\bar{t}, \tau], \tau)$, $([\tau, t], \tau)$. Denote the modified set by $\Omega(\mathcal{T})$. Then $\Omega(\mathcal{T})$ is a simple partition of $[c, d]$ which is δ-fine if $\mathcal{T}$ is δ-fine. The operation Ω may be used in order to prove that

(i) there exists a δ-fine simple partition of $[c, d]$ (cf. Lemma 14.4),
(ii) U is KH-integrable if and only if there exists $\Gamma(a, b)$ and if for $\varepsilon > 0$ there exists $\delta : [a, b] \to \mathbb{R}^+$ such that

$$\left\| \Gamma(a, b) - \sum_{\mathcal{T}} (U(\tau, t) - U(\tau, \bar{t})) \right\| \le \varepsilon$$

for every δ-fine simple partition $\mathcal{T}$ of $[a, b]$.

20.4. Definition. Let $\mathcal{T} = \{([\bar{t}, t], \tau)\}$, $\mathcal{R} = \{([\bar{r}, r], \rho)\}$ be partitions of $[c, d]$. The partition $\mathcal{R}$ is called a *refinement* of $\mathcal{T}$ if for every $([\bar{r}, r], \rho) \in \mathcal{R}$ there exists a $([\bar{t}, t], \tau) \in \mathcal{T}$ such that $[\bar{r}, r] \subset [\bar{t}, t]$.

20.5. Remark. Let $\mathcal{T} = \{([\bar{t}, t], \tau)\}$, $\mathcal{S} = \{([\bar{s}, s], \sigma)\}$ be partitions of $[a, b]$, $\delta : [a, b] \to \mathbb{R}^+$. Then there exists a δ-fine simple partition of $[a, b]$ which is a refinement of both $\mathcal{T}$ and $\mathcal{S}$ (by Remark 20.3 (i) there exists a δ-fine simple partition of every nondegenerate interval $[\bar{t}, t] \cap [\bar{s}, s]$).

20.6. Remark. Let $\mathcal{T} = \{([\bar{t}, t], \tau)\}$, $\mathcal{R} = \{([\bar{r}, r], \rho)\}$ be simple partitions of $[a, b]$, $\mathcal{R}$ being a refinement of $\mathcal{T}$. For $[\bar{t}, t], \tau) \in \mathcal{T}$, denote by $\mathcal{R}(\bar{t}, t)$ the set of $[\bar{r}, r], \rho) \in \mathcal{R}$ such that $[\bar{r}, r] \subset [\bar{t}, t]$. Then $\mathcal{R}(\bar{t}, t)$ is a simple partition of $[\bar{t}, t]$.

The next lemma plays a crucial role in the proof that U is integrable.

20.7. Lemma. *Let* $\varkappa > 0$ *and* $\delta : [c, d] \to \mathbb{R}^+$ *fulfil*

$$|\Psi(\tau)| \le \varkappa \quad \text{for} \quad \tau \in [a, b], \tag{20.8}$$

$$[\tau - \delta(\tau), \tau + \delta(\tau)] \cap [c, d] \subset \mathbf{B}(\varepsilon) \quad \text{if} \quad \tau \in \mathbf{B}(\varepsilon), \tag{20.9}$$

$$\left.\begin{array}{l} |\Psi(t) - \Psi(\tau)| < \varepsilon \\[2mm] \quad \text{if} \quad \tau \in \mathbf{B}(\varepsilon) \quad \text{and} \quad t \in [\tau - \delta(\tau), \tau + \delta(\tau)] \cap [c, d], \end{array}\right\} \tag{20.10}$$

$$[\tau - \delta(\tau), \tau + \delta(\tau)] \cap [c, d] \cap \mathbf{A}(\varepsilon) = \{\tau\} \quad \text{if} \quad \tau \in \mathbf{A}(\varepsilon), \tag{20.11}$$

$$\left.\begin{array}{l} |\Psi(t) - \Psi(\tau+)| \le \varepsilon \\[2mm] \quad \text{if} \quad \tau \in \mathbf{A}(\varepsilon) \quad \text{and} \quad \tau < t \le \min\{\tau + \delta(\tau), b\}, \end{array}\right\} \tag{20.12}$$

$$\left.\begin{array}{l} |\Psi(\tau-) - \Psi(t)| \le \varepsilon \\[2mm] \quad \text{if} \quad \tau \in \mathbf{A}(\varepsilon) \quad \text{and} \quad \max\{\tau - \delta(\tau), a\} \le t < \tau, \end{array}\right\} \tag{20.13}$$

$$\left.\begin{array}{l} \Phi(\tau + \delta(\tau)) - \Phi(\tau+) \le \dfrac{\varepsilon}{m(\varepsilon)} \quad \text{if} \quad \tau \in \mathbf{A}(\varepsilon) \quad \text{and} \quad \tau < b, \\[4mm] \Phi(\tau-) - \Phi(\tau - \delta(\tau)) \le \dfrac{\varepsilon}{m(\varepsilon)} \quad \text{if} \quad \tau \in \mathbf{A}(\varepsilon) \quad \text{and} \quad \tau > a. \end{array}\right\} \tag{20.14}$$

Let $\mathcal{T} = \{([\bar{t}, t], \tau)\}$, $\mathcal{R} = \{([\bar{r}, r], \rho)\}$ *be* δ-*fine simple partitions of* $[c, d]$, $\mathcal{R}$ *being a refinement of* $\mathcal{T}$.

Then

$$\left.\begin{array}{l} \displaystyle\sum_{\mathcal{T}} \left\| U(\tau, t) - U(\tau, \bar{t})) - \sum_{\mathcal{R}} \big(U(\rho, r) - U(\rho, \bar{r}) \big) \right\| \\[6mm] \qquad\qquad\qquad \le \varepsilon \big(\Phi(b) - \Phi(a) \big) (2 + 4\varkappa). \end{array}\right\} \tag{20.15}$$

Proof. A constant $\varkappa$ and a function δ fulfilling (20.8)–(20.14) exist since Ψ fulfils (20.1) and Φ fulfils (20.2).

The proof consists of five parts.

PART 1. Let $([\tau, t], \tau) \in \mathcal{T}$ and $\tau \in \mathbf{A}(\varepsilon)$. If $\tau < \sigma \le t$ then

$$\sigma \in \mathbf{B}(\varepsilon) \quad \text{and} \quad [\sigma - \delta(\sigma), \sigma + \delta(\sigma)] \subset \mathbf{B}(\varepsilon)$$

by (20.11). Hence there exists $([\tau, r_1], \tau) \in \mathcal{R}(\tau, t)$ and

$$U(\tau, t) - U(\tau, \tau) = U(\tau, r_1) - U(\tau, \tau) + U(\tau, t) - U(\tau, r_1). \tag{20.16}$$

On the other hand,

$$
\left.
\begin{aligned}
\sum_{\mathcal{R}(\tau,t)} & (U(\rho,r) - U(\rho,\bar{r})) \\
& = U(\tau,r_1) - U(\tau,\tau) + \sum_{\mathcal{R}^*}(U(\rho,r) - U(\rho,\bar{r})) ,
\end{aligned}
\right\} \quad (20.17)
$$

where $\mathcal{R}^* = \mathcal{R}(\tau,t) \setminus \{([\tau,r_1],\tau)\}$. By (20.16), (20.17), (20.4), (20.14)

$$
\left.
\begin{aligned}
& \left\| U(\tau,t) - U(\tau,\tau) - \sum_{\mathcal{R}(\tau,t)}(U(\rho,r) - U(\rho,\bar{r})) \right\| \\
& \leq \sum_{\mathcal{R}^*} \left\| U(\tau,r) - U(\tau,\bar{r}) - U(\rho,r) + U(\rho,\bar{r}) \right\| \\
& \leq \frac{\varepsilon}{m(\varepsilon)} \, 2 \, \varkappa \left(\Phi(b) - \Phi(a) \right)
\end{aligned}
\right\} \quad (20.18)
$$

since

$$
\left| \Psi(\rho) - \Psi(\tau) \right| \leq 2 \, \varkappa,
$$

$$
\sum_{\mathcal{R}} \left(\Phi(r) - \Phi(\bar{r}) \right) \leq \Phi(\tau + \delta(\tau)) - \Phi(\tau+) \leq \frac{\varepsilon}{m(\varepsilon)} .
$$

PART 2. Similarly, if $\tau \in \mathbf{A}(\varepsilon)$, $([\bar{t},\tau],\tau) \in \mathcal{T}$ then

$$
\left.
\begin{aligned}
& \left\| U(\tau,\tau) - U(\tau,\bar{t}) - \sum_{\mathcal{R}(\bar{t},\tau)}(U(\rho,\rho) - U(\rho,\bar{r})) \right\| \\
& \leq \frac{\varepsilon}{m(\varepsilon)} \, 2 \, \varkappa \left(\Phi(b) - \Phi(a) \right) .
\end{aligned}
\right\} \quad (20.19)
$$

PART 3. Let $([\tau,t],\tau) \in \mathcal{T}$, $\tau \in \mathbf{B}(\varepsilon)$. By (20.9), (20.10), (20.4)

$$
\left.
\begin{aligned}
& \left\| U(\tau,t) - U(\tau,\tau) - \sum_{\mathcal{R}(\tau,t)}(U(\rho,r) - U(\rho,\bar{r})) \right\| \\
& \leq \left\| \sum_{\mathcal{R}(\tau,t)}(U(\tau,r) - U(\tau,\bar{r}) - U(\rho,r) + U(\rho,\bar{r})) \right\| \\
& \leq \sum_{\mathcal{R}(\tau,t)} |\Psi(\rho) - \Psi(\tau)| \, (\Phi(r) - \Phi(\bar{r})) \leq \varepsilon \, (\Phi(t) - \Phi(\tau)) .
\end{aligned}
\right\} \quad (20.20)
$$

PART 4. Let $([\bar{t},\tau],\tau) \in \mathcal{T}$, $\tau \in \mathbf{B}(\varepsilon)$. Then similarly to PART 3

$$
\left\| U(\tau,\tau) - U(\tau,\bar{t}) - \sum_{\mathcal{R}(\bar{t},\tau)}(U(\rho,r) - U(\rho,\bar{r})) \right\| \leq \varepsilon \, (\Phi(\tau) - \Phi(\bar{t})) . \quad (20.21)
$$

PART 5.

$$\sum_{\mathcal{T}} \left[U(\tau,t) - U(\tau,\tau) + U(\tau,\tau) - U(\tau,\bar{t}) \right] - \sum_{\mathcal{R}} \left[U(\rho,r) - U(\rho,\bar{r}) \right] \left.\vphantom{\begin{array}{c}1\\2\\3\\4\\5\\6\\7\end{array}}\right\}$$

$$= \sum_1 \left[U(\tau,t) - U(\tau,\tau) - \sum_{\mathcal{R}(\tau,t)} \left(U(\rho,r) - U(\rho,\bar{r}) \right) \right]$$

$$+ \sum_2 \left[U(\tau,\tau) - U(\tau,\bar{t}) - \sum_{\mathcal{R}(\bar{t},\tau)} \left(U(\rho,r) - U(\rho,\bar{r}) \right) \right] \qquad (20.22)$$

$$+ \sum_3 \left[U(\tau,t) - U(\tau,\tau) - \sum_{\mathcal{R}(\tau,t)} \left(U(\rho,r) - U(\rho,\bar{r}) \right) \right]$$

$$+ \sum_4 \left[U(\tau,\tau) - U(\tau,\bar{t}) - \sum_{\mathcal{R}(\bar{t},\tau)} \left(U(\rho,r) - U(\rho,\bar{r}) \right) \right],$$

where $\sum_1$ runs over $([\tau,t],\tau) \in \mathcal{T}$ such that $\tau \in \mathbf{A}(\varepsilon)$, $\sum_2$ runs over $([\bar{t},\tau],\tau) \in \mathcal{T}$ such that $\tau \in \mathbf{A}(\varepsilon)$, $\sum_3$ runs over $([\tau,t],\tau) \in \mathcal{T}$ such that $\tau \in \mathbf{B}(\varepsilon)$, $\sum_4$ runs over $([\bar{t},\tau],\tau) \in \mathcal{T}$ such that $\tau \in \mathbf{B}(\varepsilon)$. (20.22) holds since $\mathbf{A}(\varepsilon) \cup \mathbf{B}(\varepsilon) = [a,b]$, $\mathbf{A}(\varepsilon) \cap \mathbf{B}(\varepsilon) = \emptyset$. The number of elements of $\mathbf{A}(\varepsilon)$ does not exceed $m(\varepsilon)$. (20.18)–(20.22) imply that (20.15) holds. The proof is complete. $\qquad\square$

20.8. Theorem. *U is* KH-*integrable,*

Proof. Let $[c,d] = [a,b]$. By Lemma 20.7, for $\varepsilon > 0$ there is $\delta : [a,b] \to \mathbb{R}^+$ such that (20.15) holds. Let $\mathcal{T} = \{([\bar{t},t],\tau)\}$, $\mathcal{S} = \{([\bar{s},s],\sigma)\}$ be δ-fine simple partitions of $[a,b]$. By Remark 20.5 then there exists a δ-fine simple partition $\mathcal{R} = \{([\bar{r},r],\rho)\}$ of $[a,b]$ which is a refinement of both $\mathcal{T}$ and $\mathcal{S}$. (20.15) implies that

$$\left\| \sum_{\mathcal{T}} \left(U(\tau,t) - U(\tau,\bar{t}) \right) - \sum_{\mathcal{T}} \sum_{\mathcal{R}(\bar{t},\tau) \cup \mathcal{R}(\tau,t)} \left(U(\rho,r) - U(\rho,\bar{r}) \right) \right\|$$

$$\leq \varepsilon \left(2 + 4\varkappa \right) \left(\Phi(b) - \Phi(a) \right),$$

$$\left\| \sum_{\mathcal{S}} \left(U(\sigma,s) - U(\sigma,\bar{s}) \right) - \sum_{\mathcal{S}} \sum_{\mathcal{R}(\bar{s},\sigma) \cup \mathcal{R}(\sigma,s)} \left(U(\sigma,r) - U(\sigma,\bar{r}) \right) \right\|$$

$$\leq \varepsilon \left(2 + 4\varkappa \right) \left(\Phi(b) - \Phi(a) \right),$$

and

$$\left. \left\| \sum_{\mathcal{T}} \left(U(\tau,t) - U(\tau,\bar{t}) \right) - \sum_{\mathcal{S}} \left(U(\sigma,s) - U(\sigma,\bar{s}) \right) \right\| \atop \leq \varepsilon \left(4 + 8\varkappa \right) \left(\Phi(b) - \Phi(a) \right) \right\} \quad (20.23)$$

since

$$\sum_{\mathcal{T}} \sum_{\mathcal{R}(\bar{t},\tau) \cup \mathcal{R}(\tau,t)} \big(U(\rho,r) - U(\rho,\bar{r})\big) = \sum_{\mathcal{S}} \sum_{\mathcal{R}(\bar{s},\sigma) \cup \mathcal{R}(\sigma,s)} \big(U(\sigma,r) - U(\sigma,\bar{r})\big).$$

U is KH-integrable by (20.23). The proof is complete. □

20.9. Theorem. *Define* $u : [a,b] \to X$ *by* $u(T) = (KH) \int_a^T D_t U(\tau,t)$. *Then* U *is SKH-integrable and* u *is its primitive.*

Moreover,

$$\|u(d) - u(c)\| \leq \Phi(d) - \Phi(c) \quad for \ \ a \leq c < d \leq b. \tag{20.24}$$

Proof. Let $[c,d] = [a,b]$ and let $\delta : [a,b] \to \mathbb{R}^+$ fulfil (20.9)–(20.14). Let $\mathcal{T} = \{([\bar{t},t],\tau)\}$ be a δ-fine partition of $[a,b]$.

Let $([\bar{t},t],\tau) \in \mathcal{T}$ be given and let $\mathcal{R} = \{([\bar{r},r],\rho)$ be a δ-fine simple partition of $[\tau,t]$. If $\tau \in \mathbf{A}(\varepsilon)$ then by (20.18)

$$\left. \begin{aligned} \Big\| U(\tau,t) - U(\tau,\tau) - \sum_{\mathcal{R}} \big(U(\rho,r) - U(\rho,\bar{r})\big) \Big\| \\ \leq \frac{\varepsilon}{m(\varepsilon)} \, 2\varkappa \big(\Phi(h) - \Phi(a)\big) \end{aligned} \right\}$$

and

$$\Big\| U(\tau,t) - U(\tau,\tau) - u(t) + u(\tau) \Big\| \leq \frac{\varepsilon}{m(\varepsilon)} \, 2\,\varkappa \big(\Phi(b) - \Phi(a)\big). \tag{20.25}$$

since $\sum_{\mathcal{R}} \big(U(\rho,r) - U(\rho,\bar{r})\big)$ can be arbitrarily close to $u(t) - u(\tau)$. If $\tau \in \mathbf{B}(\varepsilon)$ then by (20.20)

$$\left. \begin{aligned} \Big\| U(\tau,t) - U(\tau,\tau) - \sum_{\mathcal{R}} \big(U(\rho,r) - U(\rho,\bar{r})\big) \Big\| \\ \leq \varepsilon \big(\Phi(t) - \Phi(\tau)\big) \end{aligned} \right\}$$

and

$$\Big\| U(\tau,t) - U(\tau,\tau) - u(t) + u(\tau) \Big\| \leq \varepsilon \big(\Phi(t) - \Phi(\tau)\big). \tag{20.26}$$

Let $([\bar{t},\tau],\tau) \in \mathcal{T}$. Then similarly

$$\left. \begin{aligned} \Big\| U(\tau,\bar{t}) - U(\tau,\tau) - u(\bar{t}) + u(\tau) \Big\| \leq \frac{\varepsilon}{m(\varepsilon)} \, 2\,\varkappa \big(\Phi(b) - \Phi(a)\big) \\ \text{for } \tau \in \mathbf{A}(\varepsilon) \end{aligned} \right\} \tag{20.27}$$

and

$$\left\|U(\tau,\bar{t})-U(\tau,\tau)-u(\bar{t})+u(\tau)\right\| \leq \varepsilon\left(\Phi(\tau)-\Phi(\bar{t})\right) \\ \text{for } \tau \in \mathbf{B}(\varepsilon). \qquad \Bigg\} \quad (20.28)$$

(20.25)–(20.28) implies that

$$\sum_{sopT} \|u(t) - u(\bar{t}) - U(\tau,t) + U(\tau, bart)\| \leq \varepsilon\,(4\,\varkappa + 1)\left(\Phi(b) - \Phi(a)\right).$$

Hence U is SKH-integrable and u is its primitive.

(20.24) is a consequence of Lemma 14.19 and (20.3). $\square$

20.10 . Remark. The limit $U(\tau,\tau+) = \lim\limits_{t\to\tau, t>\tau} U(\tau,t)$, $a \leq \tau < b$, exists since $U(\tau,.)$ is a function of bounded variation.

Chapter 21

Integration by parts

21.1. Notation. Let $[a,b] \subset \mathbb{R}$, $U : [a,b]^2 \to X$. Put

$$V(\tau,t) = U(t,t) + U(\tau,\tau) - U(\tau,t) - U(t,\tau), \quad \tau,t \in [a,b], \quad (21.1)$$

$$W(\tau,t) = U(t,\tau), \quad \tau,t \in [a,b].$$

Define

$$(\mathrm{KH}) \int_a^b \mathrm{D}_\tau U(\tau,t) = (\mathrm{KH}) \int_a^b \mathrm{D}_t W(\tau,t) \quad (21.2)$$

if the right-hand side exists (i.e. $(\mathrm{KH}) \int_a^b \mathrm{D}_\tau U(\tau,t)$ exists if for $\varepsilon > 0$ there exists $\delta : [a,b] \to \mathbb{R}^+$ such that

$$\left\| (\mathrm{KH}) \int_a^b \mathrm{D}_\tau U(\tau,t) - \sum_{\mathcal{T}^*} (U(\tau,t) - U(\bar{\tau},t)) \right\| \leq \varepsilon$$

for every δ-fine partition $\mathcal{T}^* = \{([\bar{\tau},\tau],t)\}$ of $[a,b]$).

Similarly, define

$$(\mathrm{SKH}) \int_a^b \mathrm{D}_\tau U(\tau,t) = (\mathrm{SKH}) \int_a^b \mathrm{D}_t W(\tau,t) \quad (21.3)$$

if the right-hand side exists.

For $[c,d], [e,f] \subset [a,b]$ put

$$\widehat{U}([c,d] \times [e,f]) = U(d,f) - U(d,e) - U(c,f) + U(c,e). \quad (21.4)$$

21.2. Lemma. *Let* $\bar{t}$, τ, $t \in [a,b]$. *Then*

$$\left. \begin{array}{l} V(\tau,t) - V(\tau,\bar{t}) \\ \qquad = -U(\tau,t) + U(\tau,\bar{t}) - U(t,\tau) + U(\bar{t},\tau) + U(t,t) - U(\bar{t},\bar{t}). \end{array} \right\} \quad (21.5)$$

Let $\mathcal{T} = \{([\bar{t}, t], \tau)\}$ be a partition of $[a, b]$. Then

$$\left.\begin{aligned}\sum_{\mathcal{T}} (U(\tau, t) - U(\tau, \bar{t})) + \sum_{\mathcal{T}} (U(t, \tau) - U(\bar{t}, \tau)) - \sum_{\mathcal{T}} (U(t, t) - U(\bar{t}, \bar{t})) \\ = -\sum_{\mathcal{T}} (V(\tau, t) - V(\tau, \bar{t})).\end{aligned}\right\} \quad (21.6)$$

Proof. (21.5) is a consequence of (21.1). Further, (21.6) follows by (21.5).   □

21.3. Theorem. *If two of the integrals*

$$(\text{SKH}) \int_a^b D_t U(\tau, t), \quad (\text{SKH}) \int_a^b D_\tau U(t, \tau), \quad (\text{SKH}) \int_a^b D_\tau V(\tau, t)$$

exist, then the third one exists as well and

$$\left.\begin{aligned}(\text{SKH}) \int_a^b D_t U(\tau, t) + (\text{SKH}) \int_a^b D_\tau U(\tau, t) - U(b, b) + U(a, a) \\ = -(\text{SKH}) \int_a^b D_\tau V(\tau, t).\end{aligned}\right\} \quad (21.7)$$

Proof. Let $U_i : [a, b]^2 \to X$ be SKH-integrable, $i = 1, 2$. Definition 19.2 implies that $U_1 + U_2$ is SKH-integrable and that

$$(\text{SKH}) \int_a^b D_t U_1(\tau, t) + (\text{SKH}) \int_a^b D_t U_2(t, \tau) = (\text{SKH}) \int_a^b D_t (U_1(t, t) + U_2(\tau, t)).$$

Also, $\sum_{\mathcal{T}} (U(t, t) - T(\bar{t}, \bar{t})) = U(b, b) - U(a, a)$. Hence (21.7) can be deduced by (21.6). The proof is complete.   □

21.4. Remark. $\widehat{U}$ is *additive* in the following sense:

if $[c, d], [d, e], [f, g] \subset [a, b]$ then

$$\widehat{U}([c, e] \times [f, g]) = \widehat{U}([c, d] \times [f, g]) + \widehat{U}([d, e] \times [f, g]), \quad (21.8)$$

$$\widehat{U}([f, g] \times [c, e]) = \widehat{U}([f, g] \times [c, d]) + \widehat{U}([f, g] \times [d, e]). \quad (21.9)$$

21.5. Lemma. *Let* $a \leq \sigma < r < s \leq b$, $a \leq \bar{s} < \bar{r} < \sigma \leq b$. *Then*

$$V(\sigma, s) = V(\sigma, r) + \widehat{U}([r, s] \times [\sigma, r]) + \widehat{U}([\sigma, s] \times [r, s]), \quad (21.10)$$

$$V(\bar{s}, \sigma) = V(\bar{r}, \sigma) + \widehat{U}([\bar{s}, \sigma] \times [s, \bar{r}]) + \widehat{U}([\bar{s}, \bar{r}] \times [\bar{r}, \sigma]). \quad (21.11)$$

Proof. (21.10) holds (cf. Remark 21.4) since

$$[\sigma, s]^2 = [\sigma, r]^2 \cup \big([r, s] \times [\sigma, r]\big) \cup \big([\sigma, s] \times [r, s]\big).$$

Similarly, (21.11) follows from

$$[\bar{s}, \sigma]^2 = [\bar{r}, \sigma]^2 \cup \big([\bar{s}, \sigma] \times [\bar{s}, \bar{r}]\big) \cup \big([\bar{s}, \bar{r}] \times [\bar{r}, \sigma]\big).$$

$\square$

21.6. Lemma. *Assume that*

$$\Psi : [a, b] \to \mathbb{R} \quad \text{is a regulated function}, \tag{21.12}$$

$$\Phi : [a, b] \to \mathbb{R} \quad \text{is nondecreasing}, \tag{21.13}$$

$$\|U(\tau, t) - U(\tau, \bar{t})\| \le |\Phi(t) - \Phi(\bar{t})|, \quad \tau, \bar{t}, t \in [a, b], \tag{21.14}$$

$$\left.\begin{array}{l} \|U(\tau, t) - U(\tau, \bar{t}) - U(\sigma, t) + U(\sigma, \bar{t})\| \\ \qquad \le |\Psi(\tau) - \Psi(\sigma)|\,|\Phi(t) - \Phi(\bar{t})|, \quad \tau, \sigma, \bar{t}, t \in [a, b]. \end{array}\right\} \tag{21.15}$$

Then there exists $\varkappa > 0$ such that

$$|\Psi(\tau)| \le \varkappa, \quad \tau \in [a, b]. \tag{21.16}$$

Further,

$$\left.\begin{array}{l} \|\widehat{U}([c, d] \times [e, f])\| \le |\Psi(d) - \Psi(c)|\,|\Phi(f) - \Phi(e)| \\ \qquad\qquad\qquad\qquad\qquad \text{for } [c, d], [e, f] \subset [a, b], \end{array}\right\} \tag{21.17}$$

$$V(\tau, t) = \widehat{U}([\tau, t]^2) \quad \text{for } a \le \tau < t \le b, \tag{21.18}$$

$$\left.\begin{array}{l} \|V(\sigma, s) - V(\sigma, r)\| \\ \qquad \le |\Psi(s) - \Psi(r)|\,(\Phi(r) - \Phi(\sigma)) + |\Psi(s) - \Psi(\sigma)|\,(\Phi(s) - \Phi(r)) \\ \qquad\qquad\qquad\qquad\qquad\qquad \text{for } a \le \sigma \le r \le s \le b, \end{array}\right\} \tag{21.19}$$

$$\left.\begin{array}{l} \|V(\bar{s}, \sigma) - V(\bar{r}, \sigma)\| \\ \qquad \le |\Psi(\sigma) - \Psi(s)|\,(\Phi(\bar{r}) - \Phi(s)) + |\Psi(\bar{r}) - \Psi(\bar{s})|\,(\Phi(\sigma) - \Phi(\bar{r})) \\ \qquad\qquad\qquad\qquad\qquad\qquad \text{for } a \le \bar{s} \le \bar{r} \le \sigma \le b. \end{array}\right\} \tag{21.20}$$

Moreover, the limits

$$\lim_{t\to e, t>e} V(e,t) = \Theta_{right}(e), \quad a \le e < b, \tag{21.21}$$

$$\lim_{t\to g, t<g} V(g,t) = \Theta_{left}(g), \quad a < g \le b \tag{21.22}$$

exist and

$$\|\Theta_{right}(e)\| \le |\Psi(e+) - \Psi(e)|\,(\Phi(e+) - \Phi(e)), \tag{21.23}$$

$$\|\Theta_{left}(g)\| \le |\Psi(g) - \Psi(g-)|\,(\Phi(g) - \Phi(g-)). \tag{21.24}$$

Observe that

$$\textit{if } \Theta_{right}(e) \ne 0 \quad \textit{then } |\Psi(e+) - \Psi(e)| > 0 \ \textit{ and } \ \Phi(e+) - \Phi(e) > 0,$$
$$\textit{if } \Theta_{\ell}(e) \ne 0 \quad \textit{then } |\Psi(e) - \Psi(e-)| > 0 \ \textit{ and } \ \Phi(e) - \Phi(e-) > 0.$$

Proof.   (21.16) holds by (21.12). Further, (21.17) holds by (21.15), (21.4), (21.1) whereas (21.18) is valid by (21.4), (21.1). And (21.19) holds by (21.10), (21.17). Similarly, (21.20) follows from (21.11). It can be deduced by (21.19), (21.12), (21.13) that the limit in (21.21) exists. (21.23) holds by (21.17) since $V(\sigma,t) = \hat{U}([\sigma,t]^2)$ for $\sigma < t \le b$. (21.22) and (21.24) are proved by similar arguments. The proof is complete. □

21.7. Theorem. *Let* (21.12)–(21.15) *hold. Define* $v:[a,b] \to X$ *by*

$$\left.\begin{aligned}
&v(a) = 0, \\
&v(t) = \Theta_{right}(a) + \sum_{a<\tau<t} (\Theta_{right}(\tau) + \Theta_{left}(\tau)) + \Theta_{left}(t), \\
&\hspace{7cm} \textit{for } a < t \le b.
\end{aligned}\right\} \tag{21.25}$$

Then V is SKH-integrable and v is its primitive.

Proof.   If $\Phi(b) - \Phi(a) = 0$ then $v(t) = 0$ for $a \le t \le b$ and $V(\tau,t) = 0$ for $\tau, t \in [a,b]$ (cf. (21.12), (21.13), (21.17), (21.18)) and the theorem holds.

Assume that

$$\Phi(b) - \Phi(a) > 0. \tag{21.26}$$

Observe that the sum in (21.25) is well defined since

$$\|\Theta_{\text{right}}(\tau)\| + \|\Theta_{\text{left}}(\tau)\| \le 2\varkappa\,(\Phi(\tau+) - \Psi(\tau-))$$

by (21.23), (21.24). Let $\varepsilon > 0$. Put

$$\left.\begin{array}{l} \Lambda = \{\lambda; a \leq \lambda < b, |\Psi(\lambda+) - \Psi(\lambda)| \geq \varepsilon\} \\ \qquad \cup \{\lambda; a < \lambda \leq b, |\Psi(\lambda) - \Psi(\lambda-)| \geq \varepsilon\}. \end{array}\right\} \tag{21.27}$$

Then

$$\Lambda \text{ has a finite number } m \text{ of elements}. \tag{21.28}$$

There exists $\varkappa \in \mathbb{R}^+$ such that

$$|\Psi(\lambda)| \leq \varkappa \quad \text{for } \lambda \in [a, b]. \tag{21.29}$$

(21.28) and (21.29) are consequences of the basic properties of regulated functions. Let $\delta: [a, b] \to \mathbb{R}^+$ fulfil

$$a+\delta(a) < b, \ b-\delta(b) > a, \ a < t-\delta(t) < t+\delta(t) < b \text{ for } a < t < b, \tag{21.30}$$

$$[\lambda-\delta(\lambda), \lambda+\delta(\lambda)] \cap [a, b] \cap \Lambda = \{\lambda\} \text{ for } \lambda \in \Lambda, \tag{21.31}$$

$$\left|\Psi(t) - \Psi(\lambda+)\right| \leq \varepsilon \quad \text{for } \lambda < t \leq \lambda+\delta(\lambda), \ \lambda, t \in [a, b], \tag{21.32}$$

$$\left|\Psi(\lambda-) - \Psi(t)\right| \leq \varepsilon \quad \text{for } \lambda-\delta(\lambda) \leq t < \lambda, \ \lambda, t \in [a, b], \tag{21.33}$$

$$\Phi(\lambda-) - \Phi(\lambda-\delta(\lambda)) \leq \frac{\varepsilon}{\varkappa m} \quad \text{for } \lambda \in \Lambda, \ a < \lambda, \tag{21.34}$$

$$\Phi(\lambda + \delta(\lambda)) - \Phi(\lambda+) \leq \frac{\varepsilon}{\varkappa m} \quad \text{for } \lambda \in \Lambda, \ \lambda < b, \tag{21.35}$$

$$[a, b] \cap [\lambda - \delta(\lambda), \lambda + \delta(\lambda)] \subset [a, b] \setminus \Lambda \quad \text{for } \lambda \in [a, b] \setminus \Lambda, \tag{21.36}$$

$$\left|\Psi(t) - \Psi(\lambda-)\right| \leq \varepsilon \quad \text{if } \lambda-\delta(\lambda) \leq t < \lambda, \ \lambda, t \in [a, b], \tag{21.37}$$

$$\left|\Psi(t) - \Psi(\lambda+)\right| \leq \varepsilon \quad \text{if } \lambda < t \leq \lambda+\delta(\lambda), \ \lambda, t \in [a, b], \tag{21.38}$$

$$\Phi(\lambda-) - \Phi(t) \leq \varepsilon \quad \text{for } \lambda - \delta(\lambda) \leq t < \lambda, \ a < \lambda \leq b, \tag{21.39}$$

$$\Phi(t) - \Phi(\lambda+) \leq \varepsilon \quad \text{for } \lambda < t < \lambda+\delta(\lambda), \ a \leq \lambda < b. \tag{21.40}$$

There exists a δ which fulfils (21.30)–(21.40) since Λ is finite, Ψ is a regulated function and Φ is nondecreasing.

Let $\mathcal{S} = \{([\bar{s}, s], \sigma)\}$ be a δ-fine simple partition of $[a, b]$. Put

$$\mathcal{S}_1 = \{([\sigma, s], \sigma); |\Psi(\sigma+) - \Psi(\sigma)| \geq \varepsilon\}, \qquad (21.41)$$

$$\mathcal{S}_2 = \{([s, \sigma], \sigma); (\sigma \in [a, b] \setminus \Lambda) \text{ or } (\sigma \in \Lambda, |\Psi(\sigma+) - \Psi(\sigma)| < \varepsilon)\}, \quad (21.42)$$

$$\mathcal{S}_3 = \{([\sigma, s], \sigma); |\Psi(\sigma) - \Psi(\sigma-)| \geq \varepsilon\}, \qquad (21.43)$$

$$\mathcal{S}_4 = \{([s, \sigma], \sigma); (\sigma \in [a, b] \setminus \Lambda) \text{ or } (\sigma \in \Lambda, |\Psi(\sigma) - \Psi(\sigma-)| < \varepsilon)\}. \quad (21.44)$$

Obviously,

$$\mathcal{S} = \bigcup_{j=1}^{4} \mathcal{S}_j, \quad \mathcal{S}_i \cap \mathcal{S}_j = \emptyset \quad \text{for } i \neq j.$$

Let $([\sigma, s], s) \in \mathcal{S}_1$, $\sigma < r \leq s$. Then (cf. (21.19))

$$\|V(\sigma, s) - V(\sigma, r)\| \leq |\Psi(s) - \Psi(r)|(\Phi(r) - \Phi(\sigma)) + |\Psi(s) - \Psi(\sigma)|(\Phi(s) - \Phi(r)).$$

The limiting process for $r \to \sigma$ gives (cf. (21.21), (21.38), (21.35))

$$\left.\begin{aligned}
&\|V(\sigma, s) - \Theta_{\text{right}}(\sigma)\| \\
&\leq |\Psi(s) - \Psi(\sigma+)|(\Phi(\sigma+) - \Phi(\sigma)) + |\Psi(s) - \Psi(\sigma)|(\Phi(s) - \Phi(\sigma+)) \\
&\leq 2\varepsilon\left(\Phi(\sigma+) - \Phi(\sigma) + 2\varkappa\frac{\varepsilon}{\varkappa m}\right) \leq 2\varepsilon\left(\Phi(\sigma+) - \Phi(\sigma)\right) + \frac{2\varepsilon}{m}.
\end{aligned}\right\} \quad (21.45)$$

Moreover (cf. (21.25), (21.24), (21.23))

$$\begin{aligned}
&\|v(s) - v(\sigma) - \Theta_{\text{right}}(\sigma)\| \\
&\leq \sum_{\sigma < \lambda < s} [\|\Theta_{\text{left}}(\lambda)\| + \|\Theta_{\text{right}}(\lambda)\|] + \|\Theta_{\text{left}}(s)\| \\
&\leq \sum_{\sigma < \lambda < s} \left[|\Psi(\lambda) - \Psi(\lambda-)|(\Phi(\lambda) - \Phi(\lambda-)) + |\Psi(\lambda+) - \Psi(\lambda)|(\Phi(\lambda+) - \Phi(\lambda))\right] \\
&\qquad + |\Psi(s) - \Psi(s-)|(\Phi(s) - \Phi(s-)).
\end{aligned}$$

Hence

$$\|v(s) - v(\sigma) - \Theta_{\text{right}}(\sigma)\| \leq \varepsilon\left(\Phi(s) - \Phi(\sigma)\right) \qquad (21.46)$$

since $\sigma \in \Lambda$ and if $\sigma < \lambda \leq \sigma + \delta(\sigma)$ then $\lambda \in [a, b] \setminus \Lambda$ (cf. (21.31)) and

$$|\Psi(\lambda) - \Psi(\lambda-)| < \varepsilon, \quad |\Psi(\lambda+) - \Psi(\lambda)| < \varepsilon.$$

By (21.45), (21.47)

$$\|v(s) - v(\sigma) - V(\sigma, s)\| \le 3\,\varepsilon\,\big(\Phi(b) - \Phi(a)\big) + \frac{2\,\varepsilon}{m}$$

and

$$\sum_{\mathcal{S}_1} \|v(s) - v(\sigma) - V(\sigma, s)\| \le 3\,\varepsilon\,\big(\Phi(b) - \Phi(a)\big) + 2\,\varepsilon \qquad (21.47)$$

(cf. (21.28)).

Let $\big([\bar{s}, \sigma], \sigma\big) \in \mathcal{S}_3$, $\bar{s} \le \bar{r} < \sigma$. Then (cf. (21.1), (21.20)),

$$\|V(\sigma, \bar{s}) - V(\sigma, \bar{r})\| = \|V(\bar{s}, \sigma) - V(\bar{r}, \sigma)\|$$
$$\le \big|\Psi(\sigma) - \Psi(s)\big|\big(\Phi(\bar{r}) - \Phi(s)\big) + \big|\Psi(\bar{r}) - \Psi(\bar{s})\big|\big(\Phi(\sigma) - \Phi(\bar{r})\big).$$

The limiting process for $\bar{r} \to \sigma$ gives (cf. (21.22), (21.34), (21.37))

$$\|V(\sigma, \bar{s}) - \Theta_{\text{left}}(\sigma)\| \le 2\,\varkappa\,\frac{\varepsilon}{\varkappa\,m} + 2\,\varepsilon\,\big(\Phi(\sigma) - \Phi(\sigma-)\big). \qquad (21.48)$$

Moreover (cf. (21.25), (21.24), (21.23))

$$\|v(\sigma) - v(\bar{s}) - \Theta_{\text{left}}(\sigma)\| \le \|\Theta_{\text{right}}(\bar{s})\| + \sum_{\bar{s} < \lambda < \sigma}\big(\|\Theta_{\text{left}}(\lambda)\| + \|\Theta_{\text{right}}(\lambda)\|\big)$$

$$\le \big|\Psi(\bar{s}+) - \Psi(\bar{s})\big|\big(\Phi(\bar{s}+) - \Phi(\bar{s})\big)$$

$$+ \sum_{\bar{s} < \lambda < \sigma}\Big(\big|\Psi(\lambda) - \Psi(\lambda-)\big|\big(\Phi(\lambda) - \Phi(\lambda-)\big) + \big|\Psi(\lambda+) - \Psi(\lambda)\big|\big(\Phi(\lambda+) - \Phi(\lambda)\big)\Big).$$

Hence

$$\|v(\sigma) - v(\bar{s}) - \Theta_{\text{left}}(\sigma)\| \le \varepsilon\,\big(\Phi(\sigma) - \Phi(\bar{s})\big). \qquad (21.49)$$

By (21.47), (21.48)

$$\|v(\sigma) - v(\bar{s}) - V(\sigma, \bar{s})\| \le 3\,\varepsilon\,\big(\Phi(\sigma) - \Phi(\bar{s})\big) + \frac{2\,\varepsilon}{m}$$

and

$$\sum_{\mathcal{S}_3} \|v(\sigma) - v(\bar{s}) - V(\sigma, \bar{s})\| \le 3\,\varepsilon\,\big(\Phi(b) - \Phi(a)\big) + 2\,\varepsilon. \qquad (21.50)$$

The relations (21.47) and (21.50) give

$$\sum_{\mathcal{S}_1 \cup \mathcal{S}_3} \|v(s) - v(\bar{s}) - V(\sigma, s) + V(\sigma, \bar{s})\| \le 6\,\varepsilon\,\big(\Phi(b) - \Phi(a)\big) + 4\,\varepsilon. \qquad (21.51)$$

Let $([\sigma, s], \sigma) \in \mathcal{S}_2$. Then $\sigma < s \le \sigma + \delta(\sigma)$ and

$$\text{either} \quad \sigma \in [a,b] \setminus \Lambda \quad \text{or} \quad \sigma \in \Lambda, \; |\Psi(\lambda+) - \Psi(\lambda)| < \varepsilon. \tag{21.52}$$

In both cases (cf. (21.36), (21.36))

$$\lambda \in [a,b] \setminus \Lambda \quad \text{for} \quad \sigma < \lambda < s. \tag{21.53}$$

Hence (cf. (21.23)–(21.25))

$$\left. \begin{aligned} &\|v(s) - v(\sigma)\| \\ &\le \|\Theta_{\text{right}}(\sigma)\| + \sum_{\sigma < \lambda < s} \left(\|\Theta_{\text{right}}(\lambda)\| + \|\Theta_{\text{left}}(\lambda)\| \right) + \|\Theta_{\text{left}}(s)\| \\ &\le \varepsilon \left(\Phi(s) - \Phi(\sigma) \right) \end{aligned} \right\} \tag{21.54}$$

since

$$|\Psi(\sigma+) - \Psi(\sigma)| < \varepsilon, \; |\Psi(\lambda) - \Psi(\lambda-)| < \varepsilon,$$
$$|\Psi(\lambda+) - \Psi(\lambda)| < \varepsilon, \; |\Psi(s) - \Psi(s-)| < \varepsilon.$$

Moreover (cf. (21.4), (21.15), (21.52), (21.38))

$$\left. \begin{aligned} \|V(\sigma, s)\| &\le \|\dot{U}([\sigma, s] \times [\sigma, s]) \\ &\le \left[|\Psi(s) - \Psi(\sigma+)| + |\Psi(\sigma+) - \Psi(\sigma)| \right] \left(\Phi(s) - \Phi(\sigma) \right) \\ &\le 2\varepsilon \left(\Phi(s) - \Phi(\sigma) \right) \end{aligned} \right\} \tag{21.55}$$

since $|\Psi(s) - \Psi(\sigma+)| \le \varepsilon$ by (21.38) and $|\Psi(\sigma+) - \Psi(\sigma)| < \varepsilon$ by (21.52). The relations (21.54) and (21.55) imply that

$$\|v(s) - v(\sigma) - V(\sigma, s)\| \le 3\varepsilon \left(\Phi(s) - \Phi(\sigma) \right) \quad \text{for} \quad ([\sigma, s], \sigma) \in \mathcal{S}_2. \tag{21.56}$$

Similarly

$$\|v(\bar{s}) - v(\sigma) - V(\sigma, \bar{s})\| \le 3\varepsilon \left(\Phi(\sigma) - \Phi(\bar{s}) \right) \quad \text{for} \quad ([\bar{s}, \sigma], \sigma) \in \mathcal{S}_4. \tag{21.57}$$

By (21.56) and (21.57)

$$\left. \begin{aligned} \|v(s) - v(\bar{s}) - V(\sigma, s) + V(\sigma, \bar{s})\| &\le 3\varepsilon \left(\Phi(s) - \Phi(\bar{s}) \right) \\ &\text{for} \quad ([\bar{s}, s], \sigma) \in \mathcal{S}_2 \cup \mathcal{S}_4 \end{aligned} \right\} \tag{21.58}$$

and

$$\sum_{\mathcal{S}_2 \cup \mathcal{S}_4} \|v(s) - v(\bar{s}) - V(\sigma, s) + V(\sigma, \bar{s})\| \le 3\varepsilon \left(\Phi(b) - \Phi(a) \right). \tag{21.59}$$

Finally, (21.51) and (21.59) give

$$\sum_{S} \|v(s) - v(\bar{s}) - V(\sigma, s) + V(\sigma, \bar{s})\| \leq \varepsilon \left(9 \left(\Phi(b) - \Phi(a)\right) + 4\right).$$

V is SKH-integrable and v is its primitive. The proof is complete. □

21.8. Remark. Let (21.12)–(21.15) hold. Then $(\text{SKH}) \displaystyle\int_a^b D_\tau U(\tau, t)$ exists and

$$(\text{SKH}) \int_a^b D_t U(\tau, t)$$

$$= U(b, b) - U(a, a) - (\text{SKH}) \int_a^b D_\tau U(\tau, t) + (\text{SKH}) \int_a^b D_t V(\tau, t).$$

This is a consequence of Theorems 21.3 and 21.7.

21.9 . Remark. A similar topic was treated in [Schwabik (2001)] and [Monteiro, Tvrdý (2011)] by different methods.

Chapter 22

A variant of Gronwall inequality

22.1. Notation. Let $[a,b] \subset \mathbb{R}$ and $\Phi : [a,b] \to \mathbb{R}$. Assume that

$$\left. \begin{array}{l} \Phi \quad \text{is nondecreasing,} \\ \Phi(\tau) - \Phi(\tau-) < 1 \text{ for } a < \tau \leq b, \ \Phi(\tau+) - \Phi(\tau) < 1 \text{ for } a \leq \tau < b. \end{array} \right\} \quad (22.1)$$

For $a \leq S < T \leq b$ let $Q(S,T) = \{\sigma; \, S < \sigma < T, \, \Phi(\sigma+) - \Phi(\sigma-) > 0\}$. $Q(S,T)$ is countable and may be written as $Q(S,T) = \{\sigma_j\}_{j \in \mathbb{N}}$. The sums

$$\left. \begin{array}{c} \displaystyle\sum_{Q(S,T)} (\Phi(\sigma_j+) - \Phi(\sigma_j)), \quad \sum_{Q(S,T)} (\Phi(\sigma_j) - \Phi(\sigma_j-)), \\ \displaystyle\sum_{Q(S,T)} (\Phi(\sigma_j+) - \Phi(\sigma_j-)) \end{array} \right\} \quad (22.2)$$

are convergent and independent of the choice of the sequence $\{\sigma_j\}_{j \in \mathbb{N}}$. Therefore they are denoted by

$$\sum_{S < \sigma < T} (\Phi(\sigma+) - \Phi(\sigma)), \quad \sum_{S < \sigma < T} (\Phi(\sigma) - \Phi(\sigma-)), \quad \sum_{S < \sigma < T} (\Phi(\sigma+) - \Phi(\sigma-)).$$

Similarly we interpret some analogous cases and the products

$$\prod_{S < \sigma < T} \left(1 + \Phi(\sigma+) - \Phi(\sigma)\right), \quad \prod_{S < \sigma < T} \left(1 + \Phi(\sigma-) - \Phi(\sigma)\right), \quad \text{etc.}$$

Define

$$\left. \begin{array}{l} \Phi_J(a) = 0, \\ \Phi_J(T) = \Phi(a+) - \Phi(a) + \displaystyle\sum_{a < \tau < T} (\Phi(\tau+) - \Phi(\tau-)) + \Phi(T) - \Phi(T-) \\ \hspace{6cm} \text{for } a < T \leq b, \end{array} \right\} \quad (22.3)$$

$$\Phi_C(T) = \Phi(T) - \Phi_J(T) \quad \text{for } T \in [a,b]. \quad (22.4)$$

Φ_J is called the *jump* function, Φ_C is continuous, $\Phi_C(a) = \Phi(a)$, and both Φ_J and Φ_C are nondecreasing.

(22.3) and (22.4) imply that

$$
\left.
\begin{aligned}
\Phi(T)&-\Phi(S) \\
&= \Phi_C(T) - \Phi_C(S) + \Phi(S+) - \Phi(S) + \sum_{a<\sigma<T}(\Phi(\sigma+)-\Phi(\sigma-)) \\
&\quad +\Phi(T) - \Phi(T-), \qquad\qquad a \leq S \leq T \leq b.
\end{aligned}
\right\} \quad (22.5)
$$

Define

$$
\left.
\begin{aligned}
Z(S,S) &= 1, \\
Z(S,T) &= \exp(\Phi_C(T)-\Phi_C(S))\,(1+\Phi(S+)-\Phi(S)) \\
&\quad \times \prod_{S<\sigma<T}\left[(1+\Phi(\sigma-)-\Phi(\sigma))^{-1}\,(1+\Phi(\sigma+)-\Phi(\sigma))\right] \\
&\quad \times (1+\Phi(T-)-\Phi(T))^{-1}, \qquad a \leq S < T \leq b.
\end{aligned}
\right\} \quad (22.6)
$$

The possibly infinite product in (22.6) is convergent as we will show in Remark 22.8. Hence (cf. also (22.1)) $Z(S,T)$ is well defined.

22.2 . Remark. Let $[S,T] \subset [a,b]$, $\xi : [S,T] \to \mathbb{R}$. As in Remark 14.23, $\int_S^T \xi\, d\Phi$ or $\int_S^T \xi(\tau)\, d\Phi(\tau)$ are abbreviations for $(KH)\int_S^T D_t\xi(\tau)\,\Phi(t)$, i.e.

$$
\int_S^T \xi\, d\Phi = \int_S^T \xi(\tau)\, d\Phi(\tau) = (KH)\int_S^T D_t\xi(\tau)\,\Phi(t).
$$

Moreover,

$$
(SKH)\int_S^T D_t\xi(\tau)\,\Phi(t) = (KH)\int_S^T D_t\xi(\tau)\,\Phi(t)
$$

by Corollary 19.14.

The aim of this chapter is to prove Theorems 22.3 and 22.5.

22.3. Theorem. *Let* $\xi : [a,b] \to \mathbb{R}^+$,

$$
\xi(T) = 1 + \int_a^T \xi\, d\Phi \quad \textit{for } T \in [a,b], \tag{22.7}
$$

and let $\rho: [a, b] \to \mathbb{R}_0^+$, $\eta > 0$, and

$$\rho(T) \leq \eta + \int_a^T \rho \, d\Phi \quad for \ T \in [a, b].$$ (22.8)

Then

$$\rho(T) \leq \eta \, \xi(T) \quad for \ T \in [a, b].$$

22.4. Proposition. *The solution ξ of (22.7) is unique.*

Proof. Assume that Theorem 22.3 is valid and let (22.7) and (22.8) hold. Then by Theorem 22.3, $\zeta(T) \leq \xi(T)$ and $\xi(T) \leq \zeta(T)$ for $T \in [a, b]$. $\square$

22.5. Theorem. *Put*

$$\xi(T) = Z(a, T) \quad for \ T \in [a, b].$$ (22.9)

Then ξ is a solution of (22.7).

22.6. Remark. Let $\Phi(\tau) = \Phi(\tau-)$ for $a < \tau \leq b$. Then

$$\xi(T) \leq \exp\left(\Phi(T) - \Phi(a)\right) \quad for \ T \in [a, b].$$

This is a consequence of (22.6) since

$$Z(S, T) = \exp\left(\Phi_C(T) - \Phi_C(S)\right) \prod_{S \leq \sigma < T} \left(1 + \Phi(\sigma+) - \Phi(\sigma)\right)$$

$$\leq \exp\left(\Phi(T) - \Phi(a)\right).$$

Proof of Theorem 22.3

By (22.7) and (22.8)

$$\rho(T) - \eta\,\xi(T) \leq \int_a^T (\rho - \eta\,\xi) \, d\Phi, \quad T \in [a, b].$$ (22.10)

Put $M = \{\mu \geq a \, ; \, \int_a^T (\rho - \eta\,\xi) \, d\Phi \leq 0 \ \text{for} \ T \in [a, \mu]\}$ and $m = \sup M$. Then $a \in M$ and there exists $\nu \in [a, b]$ such that either

$$a < \nu \leq b \quad \text{and} \quad M = \{t \, ; a \leq t < \nu\},$$ (22.11)

or

$$M = [a, \nu].$$ (22.12)

Let (22.11) take place. Then (cf. (14.24), (14.25), (22.10))

$$\int_a^\nu (\rho - \eta\,\xi)\,\mathrm{d}\Phi = \int_a^{\nu-} (\rho - \eta\,\xi)\,\mathrm{d}\Phi + \int_{\nu-}^\nu (\rho - \eta\,\xi)\,\mathrm{d}\Phi$$

$$\leq 0 + (\rho(\nu) - \eta\,\xi(\nu))\,(\Phi(\nu) - \Phi(\nu-)) \leq \int_a^\nu (\rho - \eta\,\xi)\,\mathrm{d}\Phi\,(\Phi(\nu) - \Phi(\nu-)),$$

which implies that

$$\int_a^\nu (\rho - \eta\,\xi)\,\mathrm{d}\Phi \leq 0$$

since $0 = \Phi(\nu) - \Phi(\nu-) < 1$ by (22.1). Therefore (22.12) holds.

Let (22.12) take place. Assume that $\nu < b$. Then

$$\rho(\nu) - \eta\,\xi(\nu) \leq \int_a^\nu (\rho - \eta\,\nu)\,\mathrm{d}\Phi \leq 0$$

and there exists λ, $\nu < \lambda \leq b$, such that

$$\Phi(\lambda) - \Phi(\nu+) \leq \tfrac{1}{2}\,.$$

Put $\Theta = \sup\{\rho(\tau) - \eta\,\xi(\tau)\,; \nu \leq \tau \leq \lambda\}$. Then

$$\Theta \leq \sup_{\nu \leq \tau \leq \lambda} \int_a^\tau (\rho - \eta\,\xi)\,\mathrm{d}\Phi \leq \sup_{\nu \leq \tau \leq \lambda} \int_\nu^\tau (\rho - \eta\,\xi)\,\mathrm{d}\Phi$$

$$\leq \int_\nu^{\nu+} (\rho - \eta\,\xi)\,\mathrm{d}\Phi + \sup_{\nu \leq \tau \leq \lambda} \int_{\nu+}^\tau (\rho - \eta\,\xi)\,\mathrm{d}\Phi$$

$$\leq 0 + \Theta\,(\Phi(\lambda) - \Phi(\nu+)) \leq \Theta\,\tfrac{1}{2}\,.$$

Hence $\Theta \leq 0$,

$$\int_a^\lambda (\rho - \eta\,\xi)\,\mathrm{d}\Phi = \int_a^\nu (\rho - \eta\,\xi)\,\mathrm{d}\Phi + \int_\nu^\lambda (\rho - \eta\,\xi)\,\mathrm{d}\Phi \leq \Theta\,(\Phi(\lambda) - \Phi(\nu+)) \leq 0\,,$$

which contradicts the assumption $\nu < b$. Therefore $M = [a, b]$ and (22.9) holds. The proof is complete. $\qquad\square$

The proof of Theorem 22.5 is preceded by several lemmas. Some elementary inequalities are summarized in the following lemma:

22.7. Lemma. *Let $\lambda \in \mathbb{R}$, $|\lambda| \leq \frac{1}{2}$. Then*

$$|\exp(\lambda) - 1 - \lambda| \leq \lambda^2, \tag{22.13}$$

$$|(1+\lambda)^{-1} - 1 + \lambda| \leq 2\lambda^2, \tag{22.14}$$

$$|\ln(1+\lambda) - \lambda| \leq \lambda^2. \tag{22.15}$$

Proof. (22.13)–(22.15) can be deduced from the well-known expansions of functions $\exp(\lambda)$, $(1+\lambda)^{-1}$, $\ln(1+\lambda)$, where $\ln(1+\lambda)$ is the natural logarithm of $(1+\lambda)$. $\square$

22.8. Remark. The product in (22.6) is convergent since its convergence is equivalent with the convergence of the series

$$\sum_{S < \sigma < T} \left[\ln(1 + \Phi(\sigma+) - \Phi(\sigma)) - \ln(1 + \Phi(\sigma) - \Phi(\sigma-)) \right].$$

(22.15) implies that the series is convergent since the number of σ such that $\Phi(\sigma+) - \Phi(\sigma) > \frac{1}{2}$ or $\Phi(\sigma) - \Phi(\sigma-) > \frac{1}{2}$ is finite (see also (22.1)). Hence there exists $\varkappa > 1$ such that

$$1 \leq Z(S,T) \leq \varkappa \quad \text{for} \quad a \leq S < T \leq b. \tag{22.16}$$

Moreover, by (22.6)

$$Z(R,T) = Z(R,S)\, Z(S,T) \quad \text{for} \quad a \leq R \leq S < T \leq b.$$

$(Z(S,T))^{-1}$ exists by (22.16). The domain of Z may be extended to $[a,b]^2$ by putting

$$Z(S,\bar{T}) = (Z(\bar{T},S))^{-1} \quad \text{for} \quad \bar{T} < S. \tag{22.17}$$

Then

$$Z(R,T) = Z(R,S)\, Z(S,T) \quad \text{for} \quad R,S,T \in [a,b]. \tag{22.18}$$

22.9. Lemma. *Let*

$$a \leq R < S < T \leq b, \quad \Phi(T) - \Phi(S+) \leq \tfrac{1}{3}, \quad \Phi(S-) - \Phi(R) \leq \tfrac{1}{3}. \tag{22.19}$$

Then

$$Z(S,T)-Z(S,S+)$$
$$= \big(1+\Phi(S+)-\Phi(S)\big)\big(\Phi(T)-\Phi(S+)\big) + \big(1+\Phi(S+)-\Phi(S)\big)R_1\,, \quad (22.20)$$
$$\text{where } |R_1| \leq 5\big(\Phi(T)-\Phi(S+)\big)^2,$$

$$Z(S,R)-Z(S,S-)$$
$$= \big(1+\Phi(S-)-\Phi(S)\big)\big(\Phi(R)-\Phi(S-)\big) + \big(1+\Phi(S-)-\Phi(S)\big)R_2\,, \quad (22.21)$$
$$\text{where } |R_2| \leq 5\big(\Phi(S-)-\Phi(R)\big)^2.$$

Proof. By (22.6)

$$Z(S,S+) = 1 + \Phi(S+) - \Phi(S)\,, \quad (22.22)$$
$$Z(S,T) - Z(S,S+) = \big(1+\Phi(S+)-\Phi(S)\big)\big(\mathcal{A}-1\big),$$

where
$$\mathcal{A} = \exp\big(\Phi_C(T) - \Phi_C(S)\big)$$
$$\times \prod_{S<\sigma<T} \Big[\big(1+\Phi(\sigma-)-\Phi(\sigma)\big)^{-1}\big(1+\Phi(\sigma+)-\Phi(\sigma)\big)\Big]$$
$$\times \big(1+\Phi(T-)-\Phi(T)\big)^{-1},$$
$$\ln \mathcal{A} = \Phi_C(T)-\Phi_C(S) + \sum_{S<\sigma<T} \Big[\ln(1+\Phi(\sigma+)-\Phi(\sigma))$$
$$+ \sum_{S<\sigma<T} \Big[\ln(1+\Phi(\sigma+)-\Phi(\sigma) - \ln(1+\Phi(\sigma-)-\Phi(\sigma))\Big]$$
$$- \ln(1 + \Phi(T-) - \Phi(T))\Big]$$
$$= \Phi_C(T)-\Phi_C(S) + \sum_{S<\sigma<T} \big[\Phi(\sigma+)-\Phi(\sigma-)\big] + \Phi(T)-\Phi(T-) + R_3$$
$$= \Phi(T)-\Phi(S+) + R_3$$

and
$$|R_3| \leq 2\Big(\sum_{S<\sigma<T}\Big[\big(\Phi(\sigma)-\Phi(\sigma-)\big)^2 + \big(\Phi(\sigma+)-\Phi(\sigma)\big)^2\Big] + \big(\Phi(T)-\Phi(T-)\big)^2\Big)$$
$$\leq 2\big(\Phi(T)-\Phi(S+)\big)^2.$$

Hence

$$
\begin{aligned}
Z(S,T) - Z(S,S+) &= \big(1 + \Phi(S+) - \Phi(S)\big)\big[\exp\big(\Phi(T) - \Phi(S+) + R_3\big) - 1\big] \\
&= \big(1 + \Phi(S+) - \Phi(S)\big)\big[\Phi(T) - \Phi(S+) + R_1\big] \\
&= \big(1 + \Phi(S+) - \Phi(S)\big)\big(\Phi(T) - \Phi(S+)\big) + \big(1 + \Phi(S+) - \Phi(S)\big) R_1,
\end{aligned}
$$

where

$$
\begin{aligned}
|R_1| &\le |R_3| + \big(\Phi(T) - \Phi(S+) + R_3\big)^2 \\
&\le 2\big(\Phi(T) - \Phi(S+)\big)^2 + \big(\Phi(T) - \Phi(S+)\big)^2\big[1 + 4\big(\Phi(T) - \Phi(S+)\big)\big]^2 \\
&\le \big(\Phi(T) - \Phi(S+)\big)^2\big[2 + (1 + \tfrac{4}{9})^2\big] \le 5\big(\Phi(T) - \Phi(S+)\big)^2.
\end{aligned}
$$

Hence (22.20) holds. (22.21) can be proved in a similar way starting from the equations

$$
Z(S-,S) = \big(1 + \Phi(S-) - \Phi(S)\big)^{-1}, \quad Z(S,S-) = 1 + \Phi(S-) - \Phi(S), \quad (22.23)
$$

$$
\begin{aligned}
Z(S,R) &= \big(Z(R,S)\big)^{-1} \\
&= \Big(\exp\big(\Phi_C(S) - \Phi_C(R)\big)\big(1 + \Phi(R+) - \Phi(R)\big) \\
&\quad \times \prod_{R<\sigma<S}\big[(1 + \Phi(\sigma-) - \Phi(\sigma))^{-1}(1 + \Phi(\sigma+) - \Phi(\sigma))\big](1 + \Phi(S-) - \Phi(S))^{-1} \Big)^{-1}.
\end{aligned}
$$

The proof is complete. □

22.10. Lemma. *Let* (22.19) *hold. Then*

$$
\left.
\begin{aligned}
&|Z(S,T) - 1 - \Phi(T) + \Phi(S)| \\
&\quad \le \big(\Phi(S+) - \Phi(S)\big)\big(\Phi(T) - \Phi(S+)\big) + \big(1 + \Phi(S+) - \Phi(S)\big) R_1, \\
&|R_1| \le 5\big(\Phi(T) - \Phi(S+)\big)^2 \quad \text{for } a \le S < T \le b,
\end{aligned}
\right\} \quad (22.24)
$$

$$
\left.
\begin{aligned}
&|Z(S,R) - 1 - \Phi(R) + \Phi(S)| \\
&\quad \le \big(\Phi(S) - \Phi(S-)\big)\big(\Phi(S-) - \Phi(R)\big) + \big(1 + \Phi(S-) - \Phi(S)\big) R_2, \\
&|R_2| \le 5\big(\Phi(S-) - \Phi(R)\big)^2 \quad \text{for } a \le R < S \le b.
\end{aligned}
\right\} \quad (22.25)
$$

Proof. (22.24) follows by (22.22), (22.20) since

$$
\begin{aligned}
Z(S,T) - 1 - \Phi(T) + \Phi(S) &= Z(S,S+) - 1 - \Phi(S+) + \Phi(S) + Z(S,T) - Z(S,S+) - \Phi(T) + \Phi(S+).
\end{aligned}
$$

Similarly, (22.25) follows by (22.23) and (22.21). □

Proof of Theorem 22.5.

By (22.9), (22.11), (22.18)

$$\left.\begin{aligned}
&\xi(T) - \xi(S) - \xi(S)\left(\Phi(T) - \Phi(S)\right)\\
&\quad = Z(a,S)\left(Z(S,T) - 1 - \Phi(T) + \Phi(S)\right) \quad \text{for } a \le S < T \le b,
\end{aligned}\right\} \quad (22.26)$$

$$\left.\begin{aligned}
&\xi(R) - \xi(S) - \xi(S)\left(\Phi(R) - \Phi(S)\right)\\
&\quad = Z(a,S)\left(Z(S,R) - 1 - \Phi(R) + \Phi(S)\right) \quad \text{for } a \le R < S \le b.
\end{aligned}\right\} \quad (22.27)$$

Put

$$M(1) = \left\{\tau \in [a,b]; \Phi(\tau+) - \Phi(\tau) > \frac{1}{2}\right\},$$

$$N(1) = \left\{\tau \in [a,b]; \Phi(\tau) - \Phi(\tau-) > \frac{1}{2}\right\},$$

$$M(k+1) = \left\{\tau \in [a,b]; \frac{1}{2^{k+1}} < \Phi(\tau+) - \Phi(\tau) \le \frac{1}{2^k}\right\} \quad \text{for } k \in \mathbb{N},$$

$$N(k+1) = \left\{\tau \in [a,b]; \frac{1}{2^{k+1}} < \Phi(\tau) - \Phi(\tau-) \le \frac{1}{2^k}\right\} \quad \text{for } k \in \mathbb{N}.$$

The sets $M(k)$, $N(k)$ are finite. Furthermore,

$$\Phi(\tau+) = \Phi(\tau) \quad \text{for } \tau \in [a,b] \setminus \bigcup_{k \in \mathbb{N}} M(k), \quad (22.28)$$

$$\Phi(\tau-) = \Phi(\tau) \quad \text{for } \tau \in [a,b] \setminus \bigcup_{k \in \mathbb{N}} N(k). \quad (22.29)$$

Let $\varepsilon > 0$ and let $\delta:[a,b] \to \mathbb{R}^+$ fulfil the following conditions

$$\left.\begin{aligned}
&\sum_{\tau \in M(k)}\left(1 + \Phi(\tau+) - \Phi(\tau)\right)\left[\Phi(\tau+\delta(\tau)) - \Phi(\tau+) + 5\left(\Phi(\tau+\delta(\tau)) - \Phi(\tau+)\right)^2\right]\\
&\quad \le \varepsilon\, 2^{k-2} \quad \text{for } k \in \mathbb{N},
\end{aligned}\right\} \quad (22.30)$$

$$\left.\begin{aligned}
&\sum_{\tau \in N(k)}\left(1 + \Phi(\tau-) - \Phi(\tau)\right)\left[\Phi(\tau-) - \Phi(\tau-\delta(\tau)) + 5\left(\Phi(\tau-) - \Phi(\tau-\delta(\tau))\right)^2\right]\\
&\quad \le \varepsilon\, 2^{k-2} \quad \text{for } k \in \mathbb{N}.
\end{aligned}\right\} \quad (22.31)$$

(Conditions (22.30), (22.31) can be fulfilled since the sets $M(k)$, $N(k)$ are finite.) Furthermore, let

$$\left. \begin{aligned} \bigl(1+\Phi(\tau+)-\Phi(\tau)\bigr)5\bigl[\Phi(\tau+\delta(\tau))-\Phi(\tau)\bigr] &\leq \frac{\varepsilon}{\Phi(b)-\Phi(a)} \\ \text{for } \tau\in[a,b]\setminus\bigcup_{k\in\mathbb{N}}\bigl(M(k)\cup N(k)\bigr). & \end{aligned} \right\} \quad (22.32)$$

Let $\mathcal{T}=\{([\bar{t},t],\tau)\}$ be a simple δ-fine partition of $[a,b]$. Then (22.16), (22.24), (22.26) and (22.30) imply that

$$\left. \begin{aligned} &\sum_{\substack{([\bar{t},t],\tau)\in\mathcal{T}\\ \tau\in M(k)}} |\xi(t)-\xi(\tau)-\xi(\tau)\,(\Phi(t)-\Phi(\tau))| \\ &\leq \varkappa\sum_{\substack{([\tau,t],\tau)\in\mathcal{T}\\ \tau\in M(k)}} |Z(t,\tau)-1-(\Phi(t)+\Phi(\tau))| \leq \varkappa\,\varepsilon\,2^{-k-2} \quad \text{for } k\in\mathbb{N}. \end{aligned} \right\} \quad (22.33)$$

Similarly

$$\left. \sum_{\substack{([\tau,t],\tau)\in\mathcal{T}\\ \tau\in N(k)}} |\xi(\bar{t})-\xi(\tau)-\xi(\tau)\,(\Phi(\bar{t})-\Phi(\tau))| \leq \varkappa\,\varepsilon\,2^{-k-2} \quad \text{for } k\in\mathbb{N}. \right\} \quad (22.34)$$

Let $\mathcal{T}_1$ be the set of $([\tau,t],\tau)\in\mathcal{T}$ such that $\tau\in[a,b]\setminus\bigcup_{k\in\mathbb{N}} M(k)$ and let $\mathcal{T}_2$ be the set of $([\bar{t},\tau],\tau)\in\mathcal{T}$ such that $\tau\in[a,b]\setminus\bigcup_{k\in\mathbb{N}} N(k)$. Then by (22.16) and (22.24)

$$\left. \begin{aligned} \sum_{\mathcal{T}_1} |\xi(t)-\xi(\tau)-\xi(\tau)\,(\Phi(t)-\Phi(\tau))| &\leq \varkappa\sum_{\mathcal{T}_1} R_1 \\ &\leq \varkappa\sum_{\mathcal{T}_1} 5\bigl(\Phi(\tau+\delta(\tau))-\Phi(\tau)\bigr)\bigl(\Phi(t)-\Phi(\tau+)\bigr) \\ &\leq \varkappa\varepsilon\sum_{\mathcal{T}_1} \frac{\Phi(t)-\Phi(\tau+)}{\Phi(b)-\Phi(a)} \leq \varkappa\varepsilon. \end{aligned} \right\} \quad (22.35)$$

Similarly

$$\sum_{\mathcal{T}_2} |\xi(\bar{t})-\xi(\tau)-\xi(\tau)\,(\Phi(\bar{t})-\Phi(\tau))| \leq \varkappa\varepsilon. \quad (22.36)$$

(22.33)–(22.36) imply that

$$\sum_{\mathcal{T}} |\xi(t)-\xi(\bar{t})-\xi(\tau)\,(\Phi(t)-\Phi(\bar{t}))| \leq 3\,\varkappa\varepsilon.$$

Obviously $\xi(a) = 1$. Therefore ξ fulfils (22.9) The proof is complete.   □

22.11. Remark. If $\Phi(\tau-) = \Phi(\tau)$ for $a < \tau \leq b$ then

$$\xi(T) \leq \exp\left(\Phi(T) - \Phi(a)\right) \quad \text{for} \quad T \in [a, b].$$

This follows from (22.5) and (22.4) since $1 + \alpha \leq \exp(\alpha)$ for $\alpha \in \mathbb{R}^+$. Hence the estimate

$$\rho(T) \leq \eta \, \exp\left(\Phi(T) - \Phi(a)\right)$$

is valid for S from Theorem 22.3.

Chapter 23

Existence of solutions of a class of generalized ordinary differential equations

The aim of this chapter is to prove existence, uniqueness and continuous dependence theorems for the GODE

$$\frac{\mathrm{d}}{\mathrm{d}t} x = \mathrm{D}_t F(x, \tau, t).$$

In particular, the theorems are valid in the case of the equation

$$\frac{\mathrm{d}}{\mathrm{d}t} x = \mathrm{D}_t \sum_{i=1}^{k} f_i(x)\, \psi_i(\tau)\, \psi_{i+k}(t), \tag{23.1}$$

where $f_i : X \to X$ and $\mathrm{D} f_i$ fulfil a Lipschitz condition and $\psi_i : [a, b] \to \mathbb{R}$,

$$|\psi_i(\tau) - \psi_i(\sigma)| \leq |\Phi(\tau) - \Phi(\sigma)|, \quad i = 1, 2, \ldots, 2k, \quad \tau, \sigma \in [a, b].$$

23.1. Notation. Let $[a, b] \subset \mathbb{R}$, $\Phi : [a, b] \to \mathbb{R}$, $F : X \times [a, b]^2 \to X$. Let Φ be nondecreasing and continuous from the left. Assume that

$$\left. \begin{array}{c} \|F(x, \tau, t) - F(x, \tau, s)\| \leq (1 + \|x\|)\, |\Phi(t) - \Phi(s)| \\ \text{for } x \in X,\ \tau, s, t \in [a, b], \end{array} \right\} \tag{23.2}$$

$$\left. \begin{array}{c} \|\Delta_v (F(x, \tau, t) - F(x, \tau, s))\| \leq \|v\|\, |\Phi(t) - \Phi(s)| \\ \text{for } x, v \in X,\ \tau, s, t \in [a, b], \end{array} \right\} \tag{23.3}$$

$$\left. \begin{array}{c} \|F(x, \tau, t) - F(x, \tau, s) - F(x, \sigma, t) + F(x, \sigma, s)\| \\ \leq (1 + \|x\|)\, |\Phi(\tau) - \Phi(\sigma)|\, |\Phi(t) - \Phi(s)| \\ \text{for } x \in X, \tau, \sigma, s, t \in [a, b], \end{array} \right\} \tag{23.4}$$

$$\left. \begin{array}{c} \|\Delta_v \big(F(x, \tau, t) - F(x, \tau, s) - F(x, \sigma, t) + F(x, \sigma, s)\big)\| \\ \leq \|v\|\, |\Phi(\tau) - \Phi(\sigma)|\, |\Phi(t) - \Phi(s)| \\ \text{for } x, v \in X, \tau, \sigma, s, t \in [a, b]. \end{array} \right\} \tag{23.5}$$

155

Observe that the one-sided limits

$$\Phi(s+) = \lim_{t \to s, t > s} \Phi(t), \; a \leq s < b, \quad \Phi(s-) = \lim_{t \to s, t < s} \Phi(t) = \Phi(s), \; a < s \leq b,$$

$$F(x, \sigma, s+) = \lim_{t \to s, t > s} F(x, \sigma, t), \quad x \in X, \; \sigma \in [a, b], \; a \leq s < b,$$

$$F(x, \sigma, s-) = \lim_{t \to s, t < s} F(x, \sigma, t) = F(x, \sigma, s) \quad x \in X, \; \sigma \in [a, b], \; a < s \leq b, \; \text{etc.}$$

exist. Moreover,

$$\left. \begin{array}{c} \|F(x, \sigma, t) - F(x, \sigma, s+)\| \leq \left(1 + \|x\|\right) \left(\Phi(t) - \Phi(s+)\right) \\ \text{for} \; x \in X, \; \sigma \in [a, b], \; a \leq s < b \end{array} \right\} \quad (23.6)$$

and several similar inequalities are valid.

23.2. Lemma. *Let $[c, d] \subset [a, b]$ and let $w : [c, d] \to X$ be a function of bounded variation. Then*

$$\text{the integral} \; \text{(SKH)} \int_c^d \mathrm{D}_t \, F(w(\tau), \tau, t) \quad \text{exists.} \tag{23.7}$$

Define $z : [c, d] \to X$ by

$$z(T) = \text{(SKH)} \int_c^T \mathrm{D}_t \, F(w(\tau), \tau, t) \, .$$

Then

$$z \; \text{has a bounded variation and is continuous from the left.} \tag{23.8}$$

Proof. Put $U(\tau, t) = F(w(\tau), \tau, t)$ for $\tau, t \in [c, d]$. It can be deduced by (23.2)–(23.5) that U fulfils (20.3), (20.4) where Ψ, Φ are suitable nondecreasing functions (since w is bounded). (23.7) and (23.8) are true by Theorem 20.9. □

23.3. Lemma. *Let $[c, d] \subset [a, b]$, $\widehat{y} \in X$. Assume that*

$$\Phi(d) - \Phi(c+) \leq \tfrac{1}{2} \, . \tag{23.9}$$

Then there exists a function $v_{c,d} : [c, d] \to X$ such that

$$v_{c,d}(T) = \widehat{y} + \text{(SKH)} \int_c^T \mathrm{D}_t \, F(v_{c,d}(\tau), \tau, t), \quad \text{for} \; T \in [c, d], \tag{23.10}$$

$$v_{c,d} \; \text{has a bounded variation and is continuous from the left.} \tag{23.11}$$

Proof. Define

$$v_0(T) = \widehat{y} \quad \text{for} \ T \in [c,d],$$
$$\left. v_{i+1}(T) = \widehat{y} + (\text{SKH}) \int_c^T D_t \, F(v_i(\tau), \tau, t) \ \text{for} \ T \in [c,d], \ i \in \mathbb{N}_0 \, . \right\}$$ (23.12)

By Lemma 23.2 the integrals in (23.12) exist and v_i, $i \in \mathbb{N}_0$, are functions of bounded variation continuous from the left. By (23.12), Remark 20.10 and Corollary 14.18

$$v_i(c) = \widehat{y}, \quad v_i(c+) = \widehat{y}_+ \quad \text{for} \ i \in \mathbb{N}, \tag{23.13}$$

where

$$\widehat{y}_+ = \widehat{y} + F(\widehat{y}, c, c+) - F(\widehat{y}, c, c) \, . \tag{23.14}$$

By (23.12) and (23.2)

$$v_{i+1}(T) = \widehat{y}_+ + (\text{SKH}) \int_{c+}^T D_t \, F(v_i(\tau), \tau, t) \ \text{for} \ T \in [c,d], \ i \in \mathbb{N}, \tag{23.15}$$

$$v_1(T) - v_0(T) = (\text{SKH}) \int_c^T D_t \, F(\widehat{y}, \tau, t) \ \text{for} \ T \in [c,d]$$
$$\|v_1(T) - v_0(T)\| \leq (1 + \|\widehat{y}\|) \, (\Phi(T) - \Phi(c)) \ \text{for} \ T \in [c,d] \, . \tag{23.16}$$

By (23.15), (23.3) (cf. Theorem 20.9)

$$\|v_{i+2}(T) - v_{i+1}(T)\| \leq \left\| (\text{SKH}) \int_{c+}^T D_t \big[F(v_{i+1}(\tau), \tau, t) - F(v_i(\tau), \tau, t) \big] \right\|$$
$$\leq \max \{ \|v_{i+1}(\tau) - v_i(\tau)\| ; \tau \in [c,T] \} \, (\Phi(T) - \Phi(c+)) \quad \text{for} \ c < T \leq d \, .$$

Furthermore,

$$\|v_1(t) - v_0(t)\| \leq (1 + \|\widehat{y}\|) \, (\Phi(T) - \Phi(c)) \quad \text{for} \ c \leq t \leq T \leq d$$

and (by induction)

$$\left. \|v_{i+1}(T) - v_i(T)\| \leq (1 + \|\widehat{y}\|) \, (\Phi(d) - \Phi(c)) \, (\Phi(T) - \Phi(c+))^i \right\}$$
$$\left. \text{for} \ c < T \leq d, \ i \in \mathbb{N}_0 \, . \right\}$$ (23.17)

Hence (cf. (22.1))

$$\{v_i(T)\} \text{ is uniformly convergent for } c \leq T \leq d \text{ when } i \to \infty \, . \tag{23.18}$$

Put

$$v_{c,d}(T) = \lim_{i \to \infty} v_i(T), \quad T \in [c,d].$$

The limit procedure for $i \to \infty$ in (23.12) gives

$$v_{c,d}(T) = \widehat{y} + (\text{SKH}) \int_c^T D_t\, F(v_{c,d}(\tau), \tau, t), \text{ for } c \le T \le d, \qquad (23.19)$$

(23.10) holds by (23.19). (23.11) holds by Lemma 23.2. The proof is complete. $\square$

23.4. Theorem. *Let $y \in X$. Then there exists $u : [a,b] \to X$ such that*

$$u(T) = y + (\text{SKH}) \int_a^T D_t\, F(u(\tau), \tau, t) \quad \text{for } T \in [a,b]. \qquad (23.20)$$

Equivalently, u is a solution of the GODE

$$\frac{\mathrm{d}}{\mathrm{d}\,t} x = D_t\, F(x, \tau, t) \qquad (23.21)$$

on $[a,b]$ and $u(a) = y$. Moreover,

> *u has a bounded variation and is continuous from the left.* (23.22)

Proof. Let $\delta : [a,b] \to \mathbb{R}^+$ be such that

$$\Phi(\tau + \delta(\tau)) - \Phi(\tau+) \le \tfrac{1}{2} \text{ for } a \le \tau < b$$

and

$$\Phi(\tau) - \Phi(\tau - \delta(\tau)) \le \tfrac{1}{2} \text{ for } a < \tau \le b.$$

Let $\mathcal{R} = \{([r_{j-1}, r_j], \rho_j)\}$, $a = r_0 < r_1 < \ldots < r_\ell = b$, be a δ-fine partition of $[a,b]$. Let

$$\{s_i; i = 0, 1, \ldots, k\} = \{r_j\, ; j = 0, 1, \ldots, \ell\} \cup \{\rho_j\, ; j = 0, 1, \ldots, \ell\}$$

and $s_0 < s_1 < \ldots < s_k$. Then

$$\Phi(s_i) - \Phi(s_{i-1}+) \le \tfrac{1}{2} \quad \text{for } i = 1, 2, \ldots, k-1.$$

Put $y_0 = y$. By Lemma 23.3 there exists $u_1 : [s_0, s_1] \to X$ of bounded variation, continuous from the left and such that

$$u_1(T) = y_0 + (\text{SKH}) \int_{s_0}^T D_t\, F(u_1(\tau), \tau, t) \quad \text{for } s_0 \le T \le s_1.$$

Put $y_1 = u_1(s_1)$ By Lemma 23.3 there exists $u_2 \colon [s_1, s_2] \to X$ of bounded variation, continuous from the left and such that

$$u_2(T) = y_1 + (\text{SKH}) \int_{s_1}^{T} \mathrm{D}_t\, F(u_2(\tau), \tau, t) \quad \text{for } s_1 \le T \le s_2 \,.$$

By induction there exist functions $u_i \colon [s_{i-1}, s_i] \to X$, $i = 3, 4, \ldots, k$, of bounded variation, continuous from the left and $y_i \in X$, $i = 2, 3, \ldots, k-1$, such that $y_{i-1} = u_{i-1}(s_{i-1})$ and

$$u_i(T) = y_{i-1} + (\text{SKH}) \int_{s_{i-1}}^{T} \mathrm{D}_t\, F(u_i(\tau), \tau, t) \text{ for } s_{i-1} \le T \le s_i,\, i = 3, 4, \ldots, k\,.$$

Define $u \colon [a, b] \to X$ by

$$u(a) = y, \quad u(T) = u_i(T) \text{ for } s_{i-1} < T \le s_i, \quad i = 1, 2, \ldots, k\,.$$

Then

$$u(s_1) = u_1(s_1) = y_1,$$

$$u(T) = u(s_1) + (\text{SKH}) \int_{s_1}^{T} \mathrm{D}_t\, F(u(\tau), \tau, t) \quad \text{for } s_1 \le T \le s_2\,,$$

$$u(T) = y + (\text{SKH}) \int_{a}^{T} \mathrm{D}_t\, F(u(\tau), \tau, t) \quad \text{for } a \le T \le s_2\,,$$

$$u(s_2) = u_2(s_2) = y_2, \quad \ldots \text{ etc.}$$

Then u fulfils (23.20). By Lemma 23.2, the functions u_i, $i = 1, 2, \ldots, k$, have bounded variation and are continuous from the left. Therefore u fulfils (23.22). Moreover, by Definition 15.2, u is a solution of (23.21) if and only if it fulfils (23.20) and $u(a) = y$. The proof is complete. $\square$

23.5. Lemma. *Let u fulfil (23.20). Then*

u has a bounded variation and is continuous from the left, (23.23)

$$\|u(T)\| \le (\|y\| + \Phi(T) - \Phi(a)) \exp (\Phi(T) - \Phi(a)) \text{ for } T \in [a, b]\,. \quad (23.24)$$

Proof. There exists $\varkappa > 0$ such that $\|u(\tau)\| \le \varkappa$ for $\tau \in [a, b]$ since u is locally bounded by the definition of SKH-integral. Hence

$$\|u(T) - u(S)\| \le (\text{SKH}) \int_{S}^{T} \mathrm{D}_t\, F(u(\tau), \tau, t)\| \le (1 + \varkappa) (\Phi(T) - \Phi(S))$$

for $a \leq S \leq T \leq b$ and (23.23) holds.

Let $\varepsilon > 0$ and $a < T \leq b$. Then there exists $\delta : [a, T] \to \mathbb{R}^+$ such that

$$\left\| (\text{SKH}) \int_a^T D_t F(u(\tau), \tau, t) - \sum_S [F(u(\sigma), \sigma, s) - F(u(\sigma), \sigma, \bar{s})] \right\| < \varepsilon$$

for all δ-fine partitions $S = \{([\bar{s}, s], \sigma)\}$ of $[a, T]$. Hence

$$\|u(T) - u(a)\| \leq \sum_S \left(1 + \|u(\sigma)\|\right) \left(\Phi(s) - \Phi(\bar{s})\right) + \varepsilon$$

$$\leq \Phi(T) - \Phi(a) + \sum_S \|u(\sigma\| \left(\Phi(s) - \Phi(\bar{s})\right) + \varepsilon,$$

and

$$\|u(T)\| \leq \|u(a)\| + \sum_S \left(1 + \|u(\sigma)\|\right) \left(\Phi(s) - \Phi(\bar{s})\right) + \varepsilon. \tag{23.25}$$

The function $\|u\|$ has a bounded variation since u is a function of bounded variation. Hence the integral $\int_a^T \|u(\tau)\| \, d\Phi(\tau)$ exists and taking S in (23.25) sufficiently fine, we get

$$\left. \begin{array}{r} \|u(T)\| \leq \|u(a)\| + \Phi(T) - \Phi(a) + \int_a^T \|u(\tau)\| \, d\Phi(\tau) + 2\varepsilon \\ \text{for } T \in [a, b]. \end{array} \right\} \tag{23.26}$$

The relations (23.20) and (23.25) imply that

$$\|u(T)\| \leq \|y\| + \Phi(T) - \Phi(a) + \int_a^T \|u(\tau)\| \, d\Phi(\tau) \quad \text{for } T \in [a, b]$$

since $\varepsilon > 0$ can be arbitrary. Thus, by Theorem 22.3,

$$\|u(T)\| \leq \left(\|y\| + \Phi(T) - \Phi(a)\right) \xi(T),$$

where ξ is defined in (22.7) and, by Remark 22.6, fulfils

$$\|\xi(T)\| \leq \exp\left(\Phi(T) - \Phi(a)\right) \quad \text{for } T \in [a, b],$$

since Φ is continuous from the left. This implies that (23.24) is true. The proof is complete. $\qquad \square$

23.6. Lemma. *Let* $y, \overline{y} \in X$*, let* u *fulfil* (23.9) *and let* v *fulfil*

$$v(T) = \overline{y} + \int_a^T D_t F(v(\tau), \tau, t) \quad \text{for } T \in [a, b]. \tag{23.27}$$

Then

$$\|u(T) - v(T)\| \le \|y - \overline{y}\| \exp\big(\Phi(T) - \Phi(a)\big) \quad for \ T \in [a, b]. \qquad (23.28)$$

Proof. By (23.9) and (23.27),

$$u(T) - v(T) = y - \overline{y} + (SKH)\int_a^T D_t\left[F(u(\tau), \tau, t) - F(v(\tau), \tau, t)\right] \ for \ T \in [a, b].$$

Let $T \in [a, b]$. There exists $\delta: [a, b] \to \mathbb{R}^+$ such that

$$\left\|(SKH)\int_a^T D_t\left[F(u(\tau), \tau, t) - F(v(\tau), \tau, t)\right]\right.$$

$$\left. - \sum_S \left[F(u(\sigma), \sigma, s) - F(u(\sigma), \sigma, \overline{s}) - F(v(\sigma), \sigma, s) + F(v(\sigma), \sigma, \overline{s})\right]\right\| \le \varepsilon$$

whenever $S = \{([\overline{s}, s], \sigma)\}$ is a δ-fine partition of $[a, T]$. Hence

$$\|u(T) - v(T)\|$$

$$\le \|y - \overline{y}\| + \sum_S \left\|F(u(\sigma), \sigma, s) - F(u(\sigma), \sigma, \overline{s}) - F(v(\sigma), \sigma, s) + F(v(\sigma), \sigma, \overline{s})\right\| + \varepsilon$$

$$\le \|y - \overline{y}\| + \sum_S \|u(\sigma - v(\sigma)\| \left(\Phi(s) - \Phi(\overline{s})\right) + \varepsilon.$$

Consequently,

$$\|u(T) - v(T)\| \le \int_a^T \|u(\sigma) - v(\sigma)\| \, d\,\Phi(s) + \varepsilon \quad for \ T \in [a, b].$$

Now, Theorem 22.3 and Remark 22.6 yield

$$\|u(T) - v(T)\| \le (\varepsilon + \|y - \overline{y}\|) \exp\big(\Phi(T) - \Phi(a)\big) \quad for \ T \in [a, b].$$

(23.27) holds as $\varepsilon > 0$ is arbitrary. The proof is complete. □

23.7. Remark. Let u fulfil (23.9). Then u is unique. This is a consequence of Lemma 23.6.

23.8. Lemma. *Let* $y \in X$ *and let* u *fulfil* (23.20). *Let* $\overline{F}: X \times [a, b]^2 \to X$ *fulfil* (23.2)–(23.5) *with* F *replaced by* $\overline{F}$. *Assume that* $\overline{u}: [a, b] \to X$,

$$\overline{u}(T) = y + (SKH)\int_a^T D_t\overline{F}(\overline{u}(\tau), \tau, t) \quad for \ T \in [a, b], \qquad (23.29)$$

that $\Phi^*: [a, b] \to \mathbb{R}$ is nondecreasing and continuous from the left and that

$$\left. \begin{array}{c} \|F(x, \tau, t) - F(x, \tau, s) - \overline{F}(x, \tau, t) + \overline{F}(x, \tau, s)\| \\[2mm] \leq (1 + \|x\|) \left(\Phi^*(t) - \Phi^*(s)\right) \\[2mm] \textit{for} \ \ x \in X, \ s, t \in [a, b], s < t. \end{array} \right\} \qquad (23.30)$$

Then

$$\|u(T) - \overline{u}(T)\| \leq C \left(\Phi^*(b) - \Phi^*(a)\right) \textit{ for } \ T \in [a, b], \qquad (23.31)$$

where

$$C = \left(\|y\| + \Phi(b) - \Phi(a)\right) \exp \left(2 \left(\Phi(b) - \Phi(a)\right)\right).$$

Proof. By Lemma 23.5, u and $\overline{u}$ are functions of bounded variation and continuous from the left. The integral (SKH) $\displaystyle\int_a^b D_t \overline{F}(u(\tau), \tau, t)$ exists by Lemma 23.2. Therefore

$$\left. \begin{array}{l} u(T) - \overline{u}(T) = (\text{SKH}) \displaystyle\int_u^b D_t \left[F(u(\tau), \tau, t) - \overline{F}(u(\tau), \tau, t)\right] \\[4mm] \quad + (\text{SKH}) \displaystyle\int_a^b D_t \left[\overline{F}(u(\tau), \tau, t) - \overline{F}(\overline{u}(\tau), \tau, t)\right] \textit{ for } T \in [a, b]. \end{array} \right\} \qquad (23.32)$$

Let $\varepsilon > 0$ and $T \in [a, b]$. There exists $\delta: [a, b] \to \mathbb{R}^+$ such that

$$\left\| (\text{SKH}) \int_a^T D_t \left[F(u(\tau), \tau, t) - \overline{F}(u(\tau), \tau, t)\right] \right\|$$

$$\leq \sum_S \|F(u(\sigma), \sigma, s) - F(u(\sigma), \sigma, \bar{s}) - \overline{F}(u(\sigma), \sigma, s) + \overline{F}(u(\sigma), \sigma, \bar{s})\| + \varepsilon$$

$$\leq \sum_S \left(1 + \|u(\sigma)\|\right) \left(\Phi^*(s) - \Phi^*(\bar{s})\right) + \varepsilon$$

if $S = \{([\bar{s}, s], \sigma)\}$ is a δ-fine partition of $[a, T]$.

Hence

$$\left\| (\text{SKH}) \int_a^T D_t \left[F(u(\tau), \tau, t) - \overline{F}(u(\tau), \tau, t)\right] \right\| \leq \int_a^T \left(1 + \|u(\tau)\|\right) d\Phi^*(\tau) \quad (23.33)$$

since the integral on the right hand side exists and $\varepsilon > 0$ is arbitrary. By

(23.33) and (23.24),

$$\left.\begin{aligned}
&\left\| (\mathrm{SKH}) \int_a^T D_t \big[F(u(\tau),\tau,t) - \overline{F}(u(\tau),\tau,t) \big] \right\| \\
&\leq \big(\|y\| + \Phi(b) - \Phi(a) \big) \exp \big(\Phi(b) - \Phi(a) \big) \big(\Phi^*(b) - \Phi^*(a) \big).
\end{aligned}\right\} \quad (23.34)$$

$\overline{F}$ fulfils (23.3) (with F replaced by $\overline{F}$). Therefore

$$\left\| (\mathrm{SKH}) \int_a^T D_t \big[\overline{F}(u(\tau),\tau,t) - \overline{F}(\overline{u}(\tau),\tau,t) \big] \right\| \leq \int_a^T \| u(\tau) - \overline{u}(\tau) \| \, d\Phi(\tau)$$

which together with (23.32)–(23.34) implies that

$$\| u(T) - \overline{u}(T) \|$$
$$\leq \big(\|y\| + \Phi(b) - \Phi(a) \big) \exp \big(\Phi(b) - \Phi(a) \big) \big(\Phi^*(b) - \Phi^*(a) \big)$$
$$+ \int_a^T \| u(\tau) - \overline{u}(\tau) \| \, d\Phi(\tau) \quad \text{for } T \in [a,b].$$

(23.31) holds by Theorems 22.3, 22.5 and Remark 22.6. The proof is complete. □

23.9. Remark. Theorem 23.4 and Lemmas 23.6, 23.8 describe the existence, uniqueness and continuous dependence (on y and F) of solutions to the GODE (23.9). In particular, they can be applied in the case of the equation (23.1).

23.10. Lemma. *Define* $F^\circ \colon X \times [a,b] \to X$ *by*

$$F^\circ(x,t) = (\mathrm{SKH}) \int_a^t D_s F(x,\sigma,s). \quad (23.35)$$

Let $v \colon [a,b] \to X$ *be a function of bounded variation and continuous from the left. Then*

$$(\mathrm{SKH}) \int_a^T D_t F^\circ(v(\tau),t) = (\mathrm{SKH}) \int_a^T D_t F(v(\tau),\tau,t) \quad \text{for } T \in [a,b]. \quad (23.36)$$

Proof. F° is well defined by (23.35) due to Theorem 20.9 and assumptions (23.2) and (23.4). Furthermore, by (23.2) and (23.4), we have

$$\| F^\circ(x,t) - F^\circ(x,s) \| \leq \big(1 + \|x\| \big) \big(\Phi(t) - \Phi(s) \big) \quad \text{for } x \in X, \, t,s \in [a,b]. \quad (23.37)$$

Similarly,

$$\|\Delta_v\big(F^\circ(x,t) - F^\circ(x,s)\big)\| \le \|v\|\,\big(\Phi(t) - \Phi(s)\big) \quad \text{for } x, v \in X,\ t, s \in [a, b].$$

Hence F° fulfils (23.2),(23.4) (with F replaced by F°). Analogically, F° fulfils also (23.3), (23.5) since it depends only on x, t. The integrals in (23.36) exist by Lemma 23.2.

Let $\varepsilon > 0$, $a < T \le b$. There exists $\delta_1 \colon [a, b] \to \mathbb{R}^+$ such that

$$\sum_S \left\| (\text{SKH})\!\int_{\bar{s}}^{s} D_t\, F(v(\tau), \tau, t) - F(v(\sigma), \sigma, s) + F(v(\sigma), \sigma, \bar{s}) \right\| \le \varepsilon, \quad (23.38)$$

$$\sum_S \left\| (\text{SKH})\!\int_{\bar{s}}^{s} D_t\, F^\circ(v(\tau), t) - F^\circ(v(\sigma), s) + F^\circ(v(\sigma), \bar{s}) \right\| \le \varepsilon \quad (23.39)$$

whenever $S = \{([\bar{s}, s], \sigma)\}$ is δ_1-fine partition of $[a, T]$. By (23.38), (23.39), (23.35)

$$\sum_S \left\| (\text{SKH}) \int_{\bar{s}}^{s} D_t\, F(v(\tau), \tau, t) - (\text{SKH}) \int_{\bar{s}}^{s} D_t\, F^\circ(v(\tau), t) \right\|$$

$$\le \sum_S \left\| F(v(\sigma), \sigma, s) - F(v(\sigma), \sigma, \bar{s}) - F^\circ(v(\sigma), s) + F^\circ(v(\sigma), \bar{s}) \right\| + 2\varepsilon$$

$$\le \sum_S \left\| F(v(\sigma), \sigma, s) - F(v(\sigma), \sigma, \bar{s}) - (\text{SKH}) \int_{\bar{s}}^{s} D_t\, F(v(\tau), \tau, t) \right\| + 2\varepsilon.$$

By (23.38)

$$\left\| (\text{SKH}) \int_a^T D_t F(v(\tau), \tau, t) - (\text{SKH}) \int_a^T D_t F^\circ(v(\tau), t) \right\| \le 3\varepsilon \quad \text{if } a < T \le b.$$

(23.22) holds since $\varepsilon > 0$ may be arbitrary. The proof is complete. $\qquad\square$

23.11. Theorem. *Let $y \in X$, $u \colon [a, b] \to X$, $u(a) = y$. Then u is a solution of the GODE*

$$\frac{\mathrm{d}}{\mathrm{d}\,t}\, x = D_t\, F(x, \tau, t)$$

if and only if it is a solution of the GODE

$$\frac{\mathrm{d}}{\mathrm{d}\,t}\, x = D_t\, F^\circ(x, t).$$

Proof. Theorem 23.11 is a consequence of Lemma 23.10 (cf. Theorem 23.4). $\qquad\square$

Chapter 24

A convergence process as a source of discontinuities in the theory of differential equations

24.1. Motivation. Let $[a,b] \subset \mathbb{R}$, $q: X \times [a,b] \to X$, $\widehat{\varkappa} \in \mathbb{R}^+$, $p: X \to X$. Assume that

$$q \quad \text{is continuous,}$$

$$\|q(0,t)\| \leq \widehat{\varkappa}, \quad \|q(x,t) - q(y,t)\| \leq \widehat{\varkappa} \|x - y\|,$$

$$\|p(x) - p(y)\| \leq \widehat{\varkappa} \|x - y\|.$$

Then for $t, s \in [a,b]$, $s < t$, there exists a map $\mathbf{Q}(s,t): X \to X$ such that $y\,\mathbf{Q}(s,t)$ is its value at y and that the function $t \mapsto y\,\mathbf{Q}(s,t)$ is a solution of the classical equation

$$\dot{x} = q(x,t)$$

such that $y\,\mathbf{Q}(s,s) = y$. The function $y\,\mathbf{Q}(s,.)$ is uniquely defined, $\mathbf{Q}(s,t)$ is an automorphism of X and

$$y\,\mathbf{Q}(s,r)\,\mathbf{Q}(r,t) = y\,\mathbf{Q}(s,t) \quad \text{for} \quad s,r,t \in [a,b].$$

Define $H: \mathbb{R} \to \{0,1\}$ by

$$H(t) = \begin{cases} 0 & \text{for } t \leq 0, \\ 1 & \text{for } t > 0. \end{cases} \tag{24.1}$$

Let $\alpha_j, \beta, \gamma_j \in [a,b]$, $\alpha_j \leq \beta \leq \gamma_j$, $\alpha_j < \gamma_j$ for $j \in \mathbb{N}$. Assume that $\alpha_j \to \beta$, $\gamma_j \to \beta$ for $j \to \infty$. Let $u_j: [a,b] \to X$ be the solution of the classical equation

$$\dot{x} = q(x,t) + p(x)\,\frac{\chi_j(t)}{\beta_j - \alpha_j}, \quad u_j(a) = y.$$

165

χ_j being the characteristic function of $[\alpha_j, \gamma_j]$.

Let $v\,P(s,.)$ be the unique solution of

$$\dot{x} = p(x)$$

such that

$$v\,P(s,s) = v \quad \text{for} \quad s \in \mathbb{R}, \ v \in X.$$

Define $z : [a, b] \to X$ by

$$z(t) = \begin{cases} y\,\mathbf{Q}(a,t) & \text{for} \ a \leq t \leq \beta, \\ y\,\mathbf{Q}(a,\beta)(\mathrm{id} + P(0,1))\,\mathbf{Q}(\beta,t) & \text{for} \ \beta < t \leq b. \end{cases}$$

Then

$$u_j(t) \to z(t) \quad \text{for} \ j \to \infty, \ t \in [a,b], \ t \neq \beta.$$

Put

$$F(x,t) = q(x,t) + \big(x\,P(0,1)\big)\,H(t - \beta), \quad x \in X, \ t \in [a,b].$$

Then z is a solution of the GODE

$$\frac{\mathrm{d}}{\mathrm{d}\,t}\,x = \mathrm{D}_t F(t,x).$$

Observe that $F(x,.)$ is continuous from the left. The aim of this chapter is to extend the above result to a situation such that the set of discontinuities of the limit solution z may be countable.

24.2 . Notation. Let $[a,b] \subset \mathbb{R}$, $\Psi : [a,b] \to \mathbb{R}$, $Q : X \times [a,b] \to X$, $p_i : X \to X$, $\alpha_i, \beta_i, \gamma_i \in [a,b]$, $\varkappa_i \in \mathbb{R}^+$ for $i \in \mathbb{N}$. Assume that

Ψ is nondecreasing and continuous, $\qquad\qquad\qquad\qquad\qquad$ (24.2)

$a < \beta_i < b, \quad \beta_i \neq \beta_j \ \text{for} \ i \neq j, \ i,j \in \mathbb{N},$ $\qquad\qquad$ (24.3)

$$\left.\begin{aligned} \big\|Q(x,t) - Q(x,s)\big\| &\leq (1 + \|x\|)\,\big|\Psi(t) - \Psi(s)\big|, \\ &\text{for} \ x \in X, \ t,s \in [a,b], \end{aligned}\right\} \ \text{(24.4)}$$

$$\left.\begin{aligned} \big|\Delta_v\big(Q(x,t) - Q(x,s)\big)\big\| &\leq \|v\|\,\big|\Psi(t) - \Psi(s)\big|, \\ &\text{for} \ x,v \in X, \ t,s \in [a,b], \end{aligned}\right\} \ \text{(24.5)}$$

$$\sum_{i=1}^{\infty} \varkappa_i < \infty, \qquad\qquad\qquad\qquad\qquad\qquad\qquad \text{(24.6)}$$

$$\|p_i(x)\| \le \varkappa_i \left(1+\|x\|\right), \quad \|p_i(x)-p_i(y)\| \le \varkappa_i \left(\|x-y\|\right), \left.\vphantom{\begin{array}{c}a\\a\end{array}}\right\} \quad (24.7)$$
$$\text{for } x,y \in X,\, i \in \mathbb{N}.$$

24.3. Lemma. *Let $w \in X$ and $i \in \mathbb{N}$. Then there exists a unique solution $P_i(w,.)\colon \mathbb{R}_0^+ \to X$ of the classical differential equation*

$$\dot{x} = p_i(x) \tag{24.8}$$

such that $P_i(w,0)=w$.

Moreover,

$$\|P_i(w,t)\| \le \left(\|w\| + \varkappa_i t\right) \exp(\varkappa_i t) \quad \text{for } t \in \mathbb{R}_0^+, \tag{24.9}$$
$$\|P_i(w,1)-w\| \le \varkappa_i \left(\|w\|+1\right) \left[\exp(\varkappa_i)-1\right]. \tag{24.10}$$

Let $\overline{w} \in X$. Then

$$\|P_i(w,t)-P_i(\overline{w},t)\| \le \|w-\overline{w}\| \exp(\varkappa_i t) \quad \text{for } t \in \mathbb{R}_0^+, \tag{24.11}$$
$$\|P_i(w,1)-w-P_i(\overline{w},1)+\overline{w}\| \le \|w-\overline{w}\| \left[\exp(\varkappa_i)-1\right]. \tag{24.12}$$

Sketch of the proof. By standard methods the solution $P_i(w,.)$ exists for $0 \le t < \alpha$, where $\alpha > 0$, and fulfils

$$P_i(w,t) = w + \int_0^t p_i(P_i(w,\tau))\, d\tau \quad \text{for } 0 \le t < \alpha.$$

By (24.7),

$$\|P_i(w,t)\| \le \|w\| + \int_0^t \varkappa_i \left(1+\|P_i(w,\tau)\|\right) d\tau \quad \text{for } 0 \le t < \alpha.$$

By Gronwall inequality (Lemma A.1)

$$\|P_i(w,t)\| \le \left(\|w\| + \varkappa_i t\right) \exp(\varkappa_i t) \quad \text{for } 0 \le t < \alpha$$

which implies that $P_i(w,t)$ is defined on $\mathbb{R}_0^+$ and (24.9) holds. Further,

$$P_i(w,t)-w = \int_0^t p_i(P_i(w,\tau)-w)\, d\tau + \int_0^t \left[p_i(P_i(w,\tau))-p_i(P_i(w,\tau)-w)\right] d\tau$$
$$\text{for } 0 \le t < \infty.$$

By (24.7),

$$\|P_i(w,t)-w\| \le \int_0^t \varkappa_i \left(1+\|P_i(w,\tau)-w\|\right) d\tau + \int_0^t \varkappa_i \|w\|\, d\tau \quad \text{for } 0 \le t < \alpha.$$

(24.10) holds by Gronwall inequality. Similarly,

$$\|P_i(w,t) - P_i(\overline{w},t)\| \leq \|w - \overline{w}\| + \left\| \int_0^t \left[p_i(P_i(w,\tau)) - p_i(P_i(\overline{w},\tau)) \right] d\tau \right\|$$

$$\leq \|w - \overline{w}\| + \int_0^t \varkappa_i \|P_i(w,\tau)) - P_i(\overline{w},\tau)\| d\tau \quad \text{for } 0 \leq t < \alpha.$$

and (24.11) is correct by Gronwall inequality. Finally,

$$\|P_i(w,t) - w - P_i(\overline{w},t) + \overline{w}\| \leq \left\| \int_0^t \left[p_i(P_i(w,\tau)) - p_i(P_i(\overline{w},\tau)) \right] d\tau \right\|$$

$$\leq \varkappa_i \int_0^t \|P_i(w,\tau) - P_i(\overline{w},\tau)\| d\tau \quad \text{for } t \in \mathbb{R}^+$$

and (24.12) holds by (24.11). □

24.4. Theorem. *Let* $y \in X$. *Put* (*cf.* (24.1))

$$\left. \begin{array}{l} F_1(x,t) = Q(x,t) + \sum_{i=1}^{\infty} \left(P_i(x,1) - x \right) H(t - \beta_i) \\ \qquad\qquad\qquad\qquad\qquad \text{for } x \in X, t \in [a,b], \end{array} \right\} \quad (24.13)$$

$$\left. \begin{array}{l} \Phi_1(t) = \Psi(t) + \sum_{i=1}^{\infty} \left(\exp(\varkappa_i) - 1 \right) H(t - \beta_i) \\ \qquad\qquad\qquad\qquad \text{for } t \in [a,b]. \end{array} \right\} \quad (24.14)$$

Then

$$\left. \begin{array}{l} \|F_1(x,t) - F_1(x,s)\| \leq (1 + \|x\|)\, |\Phi_1(t) - \Phi_1(s)| \\ \qquad\qquad\qquad\qquad \text{for } x \in X, t, s \in [a,b], \end{array} \right\} \quad (24.15)$$

$$\left. \begin{array}{l} \|\Delta_v\big(F_1(x,t) - F_1(x,s)\big)\| \leq \|v\|\, |\Phi_1(t) - \Phi_1(s)| \\ \qquad\qquad\qquad\qquad \text{for } x, v \in X, t, s \in [a,b] \end{array} \right\} \quad (24.16)$$

and there exists $u_1 \colon [a,b] \to X$ *fulfilling*

$$u_1(T) = y + (\text{SKH}) \int_a^T D_t F_1(u_1(\tau), t) \quad \text{for } T \in [a,b]. \quad (24.17)$$

Moreover,

u_1 *is a function of bounded variation continuous from the left,* (24.18)

u_1 *is unique,* (24.19)

$\|u_1(T)\| \leq \big(\|y\| + \Phi_1(T) - \Phi_1(a)\big) \exp\big(\Phi_1(T) - \Phi_1(a)\big)$ *for* $T \subset [a,b]$. (24.20)

Proof. $\sum_{i=1}^{\infty} [\exp(\varkappa_i) - 1] < \infty$ by (24.6). Hence F_1, Φ_1 are well defined and fulfil (23.2), (23.3). Conditions (23.4), (23.5) are irrelevant since F_1 does not depend on τ. By Theorem 23.4 there exists a solution $u_1 \colon [a, b] \to X$ of (24.17), $u_1(a) = y$. It fulfils (24.18) and is unique by Lemma 23.6. Moreover, (24.20) holds by Lemma 23.5. $\qquad \square$

24.5. Notation. Let $\zeta > 0$, $a < \alpha_i < \beta_i < \gamma_i < b$ for $i \in \mathbb{N}$. There exists $k \in \mathbb{N}$ such that

$$\sum_{i=k+1}^{\infty} (\exp(\varkappa_i) - 1) \leq \zeta \tag{24.21}$$

and a permutation m of $\{1, 2, \ldots, k\}$ such that

$$\beta_{m(1)} < \beta_{m(2)} < \ldots < \beta_{m(k)}, \tag{24.22}$$

$$a < \alpha_{m(1)} < \beta_{m(1)} < \gamma_{m(1)} < \ldots < \alpha_{m(k)} < \beta_{m(k)} < \gamma_{m(k)} < b. \tag{24.23}$$

Assume that

$$\sum_{i=1}^{k} (\gamma_i - \alpha_i) \leq \zeta, \tag{24.24}$$

$$\sum_{i=1}^{k} \big(\Psi(\gamma_i) - \Psi(\alpha_i)\big) \leq \zeta, \quad \sum_{i=k+1}^{\infty} \big(\Psi(\gamma_i) - \Psi(\alpha_i)\big) \leq \zeta. \tag{24.25}$$

Define $Q_M \colon X \times [a, b] \to X$ and $\Psi^* \colon [a, b] \to \mathbb{R}$ as follows:

$Q_M(x, t) = Q(x, t), \qquad \Psi^*(t) = 0 \qquad\qquad\qquad \text{if } a \leq t \leq \alpha_{m(1)},$

$Q_M(x, t) = Q_M(x, \alpha_{m(1)}), \quad \Psi^*(t) = \Psi(t) - \Psi(\alpha_{m(1)}) \quad \text{if } \alpha_{m(1)} \leq t \leq \gamma_{m(1)},$

$Q_M(x, t) = Q_M(x, \gamma_{m(1)}) + Q(x, t) - Q(x, \gamma_{m(1)}), \quad \Psi^*(t) = \Psi^*(\gamma_{m(1)})$

$$\text{if } \gamma_{m(1)} \leq t \leq \alpha_{m(2)},$$

$Q_M(x, t) = Q_M(x, \alpha_{m(2)}), \quad \Psi^*(t) = \Psi^*(\alpha_{m(2)}) + \Psi(t) - \Psi(\alpha_{m(2)})$

$$\text{if } \alpha_{m(2)} \leq t \leq \gamma_{m(2)},$$

$Q_M(x, t) = Q_M(x, \gamma_{m(2)}) + Q(x, t) - Q(x, \gamma_{m(2)}), \quad \Psi^*(t) = \Psi^*(\gamma_{m(2)})$

$$\text{if } \gamma_{m(2)} \leq t \leq \alpha_{m(3)},$$

$\ldots \qquad\qquad \ldots \qquad\qquad\qquad \ldots$

$$Q_M(x,t) = Q(x, \gamma_{m(k)}) + Q(x,t) - Q(x, \gamma_{m(k)}), \quad \Psi^*(t) = \Psi^*(\gamma_{m(k)})$$

$$\text{if } \gamma_{m(k)} \leq t \leq b.$$

Put

$$F_2(x,t) = Q_M(x,t) + \sum_{i=k+1}^{\infty} \left(P_i(x,1) - x \right) H(t - \beta_i) \left. \right\} \quad (24.26)$$
$$\text{for } x \in X, t \in [a,b],$$

$$\Phi_1^*(t) = \Psi^*(t) + \sum_{i=k+1}^{\infty} \left(\exp(\varkappa_i) - 1 \right) H(t - \beta_i) \left. \right\} \quad (24.27)$$
$$\text{for } x \in X, t \in [a,b].$$

24.6. Remark. The number k and the permutation m having the properties mentioned in Notation 24.5 exist by (24.6) and (24.3), respectively. Moreover (cf. (24.25)),

$$\Psi^* \quad \text{is nondecreasing and continuous,,} \quad \Psi^*(a) = 0, \quad \Psi^*(b) < \zeta, \quad (24.28)$$

Φ_1^* is nondecreasing and continuous from the left,
$$\left. \right\} \quad (24.29)$$
$$\Phi_1^*(b) - \Phi_1^*(a) < 2\zeta.$$

By (24.10) and (24.12),

$$\|P_i(x,1) - x\| \leq \left(1 + \|x\| \right) \left(\exp(\varkappa_i) - 1 \right), \quad (24.30)$$
$$\|\Delta_v \left(P_1(x,1) - x \right)\| \leq \|v\| \left(\exp(\varkappa_i) - 1 \right) \quad (24.31)$$

for $x \in X$. Hence

$$\|F_2(x,t) - F_2(x,s)\| \leq \left(1 + \|x\| \right) |\Phi_1(t) - \Phi_1(s)| \left. \right\} \quad (24.32)$$
$$\text{for } x \in X, t,s \in [a,b],$$

$$\|\Delta_v \left(F_2(x,t) - F_2(x,s) \right)\| \leq \|v\| |\Phi_1(t) - \Phi_1(s)| \left. \right\} \quad (24.33)$$
$$\text{for } x, v \in X, t,s \in [a,b].$$

24.7. Theorem. *Let* $y \in X$. *Then there exists* $u_2 \colon [a,b] \to X$ *fulfilling*

$$u_2(T) = y + (\mathrm{SKII}) \int_a^T D_t F_2(u_2(\tau), t) \quad \text{for} \quad T \in [a,b]. \quad (24.34)$$

Moreover,

u_2 *has a bounded variation and is continuous from the left,*   (24.35)

u_2 *is unique,*   (24.36)

$$\left. \begin{array}{c} \|u_2(T)\| \le \big(\|y\| + \Phi_1(T) - \Phi_1(a)\big) \, \exp\big(\Phi_1(T) - \Phi_1(a)\big) \\[2mm] \text{for } \ T \in [a,b]. \end{array} \right\} \quad (24.37)$$

The **proof** is omitted since it follows the same arguments as the proof of Theorem 24.4 (cf. (24.32), (24.33)).   □

24.8. Theorem. $\|u_1(T) - u_2(T)\| \le 2\,C_1\,\zeta,$ *where*

$$C_1 = \big(\|y\| + \Phi_1(b) - \Phi_1(a)\big) \, \exp\big(2\,(\Phi_1(b) - \Phi_1(a))\big).$$

Hint: Theorem 24.8 is a consequence of Lemma 23.8, where $F = F_1$, $\overline{F} = F_2$, $\Phi^* = \Phi_1^*$ and $C = C_1$ (cf. (24.29)) since

$$\|F_1(x,t) - F_1(x,s) - F_2(x,t) + F_2(x,t)\| \le \big(1 + \|x\|\big)\big(\Phi_1^*(t) - \Phi_1^*(s)\big)$$
$$\text{for } x \in X,\ s,\, t \in [a,b],\ s < t,$$

$\Phi_1^*(b) - \Phi_1^*(a) \le 2\,\zeta.$

24.9. Lemma. *For* $i \in \mathbb{N}$ *put*

$$\omega_i(t) = \begin{cases} 0 & \text{if } a \le t \le \alpha_i, \\ (\gamma_i - \alpha_i)^{-1} & \text{if } \alpha_i < t < \gamma_i, \\ 0 & \text{if } \gamma_i \le t \le b, \end{cases} \quad (24.38)$$

$$\Omega_i(t) = \int_a^t \omega_i(\tau)\,\mathrm{d}\tau \quad \text{for } t \in [a,b], \quad (24.39)$$

$$\Phi_3(t) = \Psi(t) + \sum_{i=1}^{\infty} \big(\exp(\varkappa_i) - 1\big)\,\Omega_i(t) \quad \text{for } t \in [a,b], \quad (24.40)$$

$$F_3(x,t) = Q(x,t) + \sum_{i=1}^{\infty} p_i(x)\,\Omega_i(t) \quad \text{for } x \in X, t \in [a,b], \quad (24.41)$$

$$F_4(x,t) = Q_M(x,t) + \sum_{i=1}^{k} p_i(x)\,\Omega_i(t) \quad \text{for } x \in X, t \in [a,b], \quad (24.42)$$

$$\Phi_3^*(t) = \Psi^*(t) + \sum_{i=k+1}^{\infty} \big(\exp(\varkappa_i) - 1\big) \int_a^t \omega_i(s)\,\mathrm{d}s \quad \text{for } t \in [a,b]. \quad (24.43)$$

(For Q_M, $\Psi^*(t)$ see Notation 24.5, for k see (24.21).)

Then

$$\|F_3(x,t) - F_3(x,s)\| \leq (1 + \|x\|)\,|\Phi_3(t) - \Phi_3(s)| \left. \atop \quad for\ x \in X,\, t,s \in [a,b]\,, \right\} \quad (24.44)$$

$$\|\Delta_v\big(F_3(x,t) - F_3(x,s)\big)\| \leq \|v\|\,|\Phi_3(t) - \Phi_3(s)| \left. \atop \quad for\ x,v \in X,\, t,s \in [a,b]\,, \right\} \quad (24.45)$$

$$\|F_4(x,t) - F_4(x,s)\| \leq (1 + \|x\|)\,|\Phi_3(t) - \Phi_3(s)| \left. \atop \quad for\ x \in X,\, t,s \in [a,b]\,, \right\} \quad (24.46)$$

$$\|\Delta_v\big(F_4(x,t) - F_4(x,s)\big)\| \leq \|v\|\,|\Phi_3(t) - \Phi_3(s)| \left. \atop \quad for\ x,v \in X,\, t,s \in [a,b]\,, \right\} \quad (24.47)$$

$$\Phi_3(b) - \Phi_3(a) = \Phi_1(b) - \Phi_1(a)\,. \quad (24.48)$$

Hint: The series in (24.40), (24.41) are convergent by (24.7) since $0 \leq \Omega_i(t) \leq 1$. (24.44)–(24.47) are valid by (24.4), (24.5) and (24.37). Finally, (24.48) holds since $\Omega_i(b) = 1$ for $i \in \mathbb{N}$. ⊓

24.10. Theorem. *Let* $y \in X$. *Then there exists* $u_3 : [a,b] \to X$ *fulfilling*

$$u_3(T) = y + (\mathrm{SKH}) \int_a^T D_t F_3(u_3(\tau), t) \quad for\ T \in [a,b]\,. \quad (24.49)$$

Moreover,

u_3 *has a bounded variation and is continuous from the left,*   (24.50)

u_3 *is unique,*   (24.51)

$$\|u_3(T)\| \leq \big(\|y\| + \Phi_3(T) - \Phi_3(a)\big)\,\exp\big(\Phi_3(T) - \Phi_3(a)\big) \left. \atop \quad for\ T \in [a,b]\,. \right\} \quad (24.52)$$

The **proof** is omitted since it follows the same arguments as the proof of Theorem 24.4. □

    Similarly, the proofs of Theorems 24.11, 24.12 are omitted (cf. the proofs of Theorems 24.4 and 24.8 together with (24.48) and observe that $\Phi_3^*(b) - \Phi_3^*(a) \leq 2\zeta$ and $\Phi_3(b) - \Phi_3(a) = \Phi_1(b) - \Phi_1(a)$).

24.11. Theorem. *Let* $y \in X$. *Then there exists* $u_4 : [a, b] \to X$ *fulfilling*

$$u_4(T) = y + (\text{SKH}) \int_a^T D_t F_4(u_4(\tau), t) \quad \text{for} \quad T \in [a, b]. \qquad (24.53)$$

Moreover,

    u_4 *has a bounded variation and is continuous from the left,*   (24.54)

    u_4 *is unique,*   (24.55)

$$\left. \begin{aligned} \|u_4(T)\| \leq \big(\|y\| + \Phi_1(T) - \Phi_1(a) \big) \, \exp \big(\Phi_1(T) - \Phi_1(a) \big) \\ \textit{for} \quad T \in [a, b]. \end{aligned} \right\} \quad (24.56)$$

24.12. Theorem. $\|u_3(T) - u_4(T)\| \leq 2 \, C_1 \, \zeta.$

24.13. Lemma.

$$\left. \begin{aligned} u_4(t) = u_3(t) \quad \textit{for} \quad t \in \mathcal{A}, \quad \textit{where} \\ \mathcal{A} = [a, \alpha_{m(1)}] \cup [\gamma_{m(1)}, \alpha_{m(2)}] \cup \ldots \cup [\gamma_{m(k-1)}, \alpha_{m(k)}] \cup [\gamma_{m(k)}, b] \, , \end{aligned} \right\} \quad (24.57)$$

$$|[a, b] \setminus \mathcal{A}| < \zeta. \qquad (24.58)$$

Proof.   Observe that (cf. (24.26), (24.42))

$$\left. \begin{aligned} F_4(x, t) - F_4(x, s) = F_3(x, t) - F_3(x, s) \quad \text{for} \\ t, s \in [a, \alpha_{m(1)}] \text{ or } t, s \in [\gamma_{m(j)}, \alpha_{m(j+1)}], \; j = 1, 2, \ldots, k-1, \\ \text{or } t, s \in [\gamma_{m(k)}, b] \, . \end{aligned} \right\} \quad (24.59)$$

By (24.26),

$$u_3(\gamma_{m(j)}) = P_j(\alpha_{m(j)}, 1) \quad \text{for} \quad j = 1, 2, \ldots, k \qquad (24.60)$$

and, by (24.42),

$$u_4(t) = u_4(\alpha_{m(j)}) + (\text{SKH}) \int_{\alpha_{m(j)}}^t D_s F_4(u_4(\sigma), s)$$

$$= u_4(\alpha_{m(j)}) + (\text{SKH}) \int_{\alpha_{m(j)}}^t p_j(u_4(\sigma)) \left[\gamma_{m(j)} - \alpha_{m(j)} \right]^{-1} \mathrm{d}\sigma,$$

for $j = 1, 2, \ldots, k$. Lemma 24.3 gives

$$u_4(t) = P_j(u_4(\alpha_j), t) \quad \text{for} \quad \alpha_{m(j)} \leq t \leq \gamma_{m(j)}, \, j = 1, 2, \ldots, k \, .$$

In particular,

$$u_4(\gamma_{m(j)}) = P_j(u_4(\alpha_j), \gamma_{m(j)}) \quad \text{for} \quad j = 1, 2, \ldots, k.\qquad(24.61)$$

(24.59)–(24.61) imply successively that

$$u_4(t) = u_3(t) \text{ for } t \in [a, \alpha_{m(1)}], \qquad u_4(\gamma_{m(1)}) = u_2(\gamma_{m(1)}),$$
$$u_4(t) = u_3(t) \text{ for } t \in [\gamma_{m(1)}, \alpha_{m(2)}], \qquad u_4(\gamma_{m(2)}) = u_2(\gamma_{m(2)}),$$

$$\cdots \qquad\qquad \cdots \qquad\qquad \cdots$$

$$u_4(t) = u_3(t) \text{ for } t \in [\gamma_{m(k-1)}, \alpha_{m(k)}], \quad u_4(\gamma_{m(k)}) = u_2(\gamma_{m(k)}),$$
$$u_4(t) = u_3(t) \text{ for } t \in [\gamma_{m(k)}, b].$$

Hence (24.57) is true. (24.58) is a consequence of (24.24). The proof is complete. $\qquad\qquad\qquad\qquad\qquad\qquad\qquad\qquad\qquad\qquad\qquad\square$

24.14. Theorem. *There is $\mathcal{A} \subset [a, b]$ such that*

$$\|u_1(T) - u_3(T)\| \le 2C_1\zeta \quad \text{for} \quad T \in \mathcal{A} \quad \text{and} \quad |[a, b] \setminus \mathcal{A}| < \zeta.\qquad(24.62)$$

Proof follows from Theorems 24.8, 24.12 and Lemma 24.13. $\qquad\qquad\square$

24.15. Remark. Let there exist sequences $(\alpha_{i,j}, j \in \mathbb{N})$, $(\gamma_{i,j}, j \in \mathbb{N})$ for $i \in \mathbb{N}$ such that

$$a < \alpha_{i,j} \le \beta_i \le \gamma_{i,j} < b, \ \alpha_{i,j} < \gamma_{i,j},$$

and

$$\alpha_{i,j} \to \beta_i, \ \gamma_{i,j} \to \beta_i \quad \text{for } j \to \infty \text{ and } i \in \mathbb{N}.$$

For $\ell \in \mathbb{N}$ put $\zeta_\ell = 2^{-\ell}$. There exist $k = k(\ell)$ such that (24.21) holds with $\zeta = \zeta_\ell$, the permutation $m = m(\ell)$ such that (24.22) holds and $\alpha_{m(1)}, \gamma_{m(1)}, \ldots, \alpha_{m(k)}, \gamma_{m(k)}$ such that (24.23)–(24.25) hold. Define $Q_M = Q_{M,\ell}$, $\Psi^* = \Psi_\ell^*$ in the same way as in Notation 24.5, $F_2 = F_{2,\ell}$ by (24.26), Φ_1^* by (24.27), $\omega_i = \omega_{i,\ell}$, $\Omega_i = \Omega_{i,\ell}$, $\Phi_3 = \Phi_{3,\ell}$ by (24.38)–(24.40), F_3, F_4 by (24.41), (24.42), $\Phi_3^* = \Phi_{3,\ell}^*$ by (24.43). Then all the objects introduced from Notation 24.5 till Lemma 24.13 — in particular, $\alpha_i = \alpha_{i,\ell}$, $\gamma_i = \gamma_{i,\ell}$ for $i \in \mathbb{N}$, $k = k(\ell)$, the permutation $m = m_\ell$, $Q_M = Q_{M,\ell}$, $\Psi^* = \Psi_\ell^*$, $\omega_i = \omega_{i,\ell}$, $\Omega_i = \Omega_{i,\ell}$, $\Phi_3 = \Phi_{3,\ell}$, $u_3 = u_{3,\ell}$, $\mathcal{A} = \mathcal{A}_\ell$ - are dependent on ℓ. But the difference $\Phi_{3,\ell}(b) - \Phi_{3,\ell}(a)$ is independent of ℓ since

$$\Phi_{3,\ell}(b) - \Phi_{3,\ell}(a) = \Phi(b) - \Phi(a)$$

(cf. (24.48)) and C_1 is also independent of ℓ (cf. Theorem 24.8). By Theorem 24.10,

$$\|u_{3,\ell}(T)\| \leq \big(\|y\| + \Phi_1(b) - \Phi_1(a)\big) \exp\big(2(\Phi_1(b) - \Phi_1(a))\big) \text{ for } T \in [a,b], \ell \in \mathbb{N},$$

i.e. functions $u_{3,\ell}$ are bounded uniformly.

By Theorem 24.14,

$$\|u_{3,\ell}(T) - u_1(T)\| \leq 4\,C_1\,2^{-\ell} \quad \text{for } T \in \mathcal{A}_\ell, \, \ell \in \mathbb{N}$$

and

$$\big|[a,b] \setminus \mathcal{A}_\ell\big| \leq 2^{-\ell} \quad \text{for } \ell \in \mathbb{N}.$$

Let $n \in \mathbb{N}$. Then

$$\|u_{3,\ell} - u_1(T)\| \leq 4\,C_1\,2^{-\ell} \quad \text{for } T \in \bigcap_{j=n}^{\infty} \mathcal{A}_\ell, \, \ell = n, n+1, n+2, \ldots$$

and

$$\Big|[a,b] \setminus \bigcap_{j=n}^{\infty} \mathcal{A}_j\Big| \leq \sum_{j=n}^{\infty} \big|[a,b] \setminus \mathcal{A}_j\big| \leq 2^{-n+1}.$$

Hence

$$u_{3,\ell} \to u_1 \quad \text{almost everywhere.}$$

The above results can be summarized:

Let $Q(.,a)$ be continuous. Then the functions Q, $F_{3,\ell}$ are continuous while F_1 The function $u_{3,\ell}$ is a solution to

$$\frac{\mathrm{d}}{\mathrm{d}t} x = D_t F_{3,\ell}(x,t), \quad u_{3,\ell}(a) = y,$$

on $[a,b]$ and u_1 is a solution to

$$\frac{\mathrm{d}}{\mathrm{d}t} x = D_t F_1(x,t), \quad u_1(a) = y.$$

on $[a,b]$. The functions $u_{3,\ell}$ are uniformly bounded and $u_{3,\ell} \to u_1$ almost everywhere as $\ell \to \infty$.

Chapter 25

A class of Strong Riemann-integrable functions

Let $[a, b] \subset \mathbb{R}$, $Q: [a, b]^2 \to X$. Assume that

$$Q \text{ is continuous}, \tag{25.1}$$

$$\|Q(\tau, t) - Q(\tau, s)\| \le \psi_2(|t - s|) \quad \text{for } t, s, \tau \in [a, b], \tag{25.2}$$

$$\left. \begin{array}{c} \|Q(\tau, t) - Q(\tau, s) - Q(\sigma, t) + Q(\sigma, s)\| \le \psi_1(|\tau - \sigma|) \, \psi_2(|t - s|) \\[2mm] \text{for } t, s, \tau, \sigma \in [a, b], \end{array} \right\} \tag{25.3}$$

ψ_1, ψ_2 being introduced in Chapter 7.

The purpose of this chapter is to prove that Q is SR-integrable on $[a, b]$. The chapters 25 and 26 are preparatory for for the which is formulated in Theorem 27.5.

25.1. Notation. Put $c = 2b - a$ and extend Q to $[a, b] \times [a, c]$ in such a way that $Q(\tau, t) = Q(\tau, b)$ if $\tau \in [a, b]$, $t \in [b, c]$, i.e.

$$Q(\tau, t) = Q(\tau, \min\{t, b\}) \quad \text{for } \tau \in [a, b], \, t \in [a, c]. \tag{25.4}$$

Then the inequalities in (25.2), (25.3) are valid if $\tau, \sigma \in [a, b]$, $t, s \in [a, c]$ and Q is continuous on its new domain. Put

$$\mathbf{Q}(S, t, T) = \begin{cases} 0 & \text{if } a \le t \le S, \\ Q(S, t) - Q(S, S) & \text{if } S < t \le T, \\ Q(S, T) - Q(S, S) & \text{if } T < t \le b, \end{cases} \tag{25.5}$$

$$\mu_i = (b-a)\, 2^{-i}, \quad \nu_i = (T-S)\, 2^{-i}, \tag{25.6}$$

$$\left.\begin{aligned}
R_i(S,t,T) &= \sum_{j=1}^{2^i} \mathbf{Q}(S+(j-1)\,\nu_i, t, S+j\,\nu_i) \\
&\quad \text{for } i\in\mathbb{N}_0,\ S\in[a,b],\ t\in[a,c],\ T\in[S,c],
\end{aligned}\right\} \tag{25.7}$$

$$P_i(t) = \sum_{j=1}^{2^i} \mathbf{Q}(a+(j-1)\,\mu_i, t, a+j\,\mu_i) \quad \text{for } i\in\mathbb{N}_0,\ t\in[a,c]. \tag{25.8}$$

Observe that

$$P_i(t) = P_i(b) \qquad \text{for } t\in[b,c],\ i\in\mathbb{N}_0, \tag{25.9}$$

$$P_i(t) = R_i(a,t,b) \quad \text{for } t\in[a,c],\ i\in\mathbb{N}_0. \tag{25.10}$$

25.2. Theorem. *The limit on the right-hand side of*

$$P(t) = \lim_{i\to\infty} P_i(t) \tag{25.11}$$

exists for each $t\in[a,b]$. Furthermore, for $\varepsilon>0$ there exists $\xi>0$ such that

$$\left.\begin{aligned}
\|P(t)-P(s)-Q(s,t)+Q(s,s)\| &\leq \varepsilon\,|t-s| \\
&\text{for } t,s\in[a,b],\ |t-s|\leq\xi.
\end{aligned}\right\} \tag{25.12}$$

Moreover,

*   *Q is SR-integrable on $[a,b]$ and P is its primitive.* $\qquad$ (25.13)

Proof. After some manipulation (25.7) gives

$$\left.\begin{aligned}
R_{i+1}(S,t,T) - R_i(S,t,T) &= \sum_{j=1}^{2^i} L_{i,j}(t) \\
&\text{for } S\in[a,b],\ t\in[a,c],\ i\in\mathbb{N}_0,
\end{aligned}\right\} \tag{25.14}$$

where

$$\left.\begin{aligned}
L_{i,j}(t) = {}&\mathbf{Q}(S+(2j-2)\,\nu_{i+1}, t, S+(2j-1)\,\nu_{i+1}) \\
&+ \mathbf{Q}(S+(2j-1)\,\nu_{i+1}, t, S+2j\,\nu_{i+1}) \\
&- \mathbf{Q}(S+(j-1)\,\nu_i, t, S+j\,\nu_i).
\end{aligned}\right\} \tag{25.15}$$

If $t \leq S + (2j-1)\nu_{i+1}$ then (cf. (25.5))

$$\mathbf{Q}(S + (2j-1)\nu_{i+1}, t, S + 2j\nu_{i+1}) = 0$$

and

$$\mathbf{Q}(S+(2j-2)\nu_{i+1}, t, S+(2j-1)\nu_{i+1}) = \mathbf{Q}(S+(j-1)\nu_i, t, S+j\nu_i).$$

Therefore

$$L_{i,j}(t) = 0 \quad \text{for} \quad a \leq t \leq S + (2j-1)\nu_{i+1}, \ i \in \mathbb{N}, \ j = 1, 2, \ldots, 2^i. \quad (25.16)$$

If $S + (2j-1)\nu_{i+1} \leq t < S + 2j\nu_{i+1}$ then

$$\left.\begin{aligned}
L_{i,j}(t) &= Q(S + (2j-2)\nu_{i+1}, S + (2j-1)\nu_{i+1}) \\
&\quad - Q(S + (2j-2)\nu_{i+1}, S + (2j-2)\nu_{i+1}) \\
&\quad + Q(S + (2j-1)\nu_{i+1}, t) \\
&\quad - Q(S + (2j-1)\nu_{i+1}, S + (2j-1)\nu_{i+1}) \\
&\quad - Q(S + (j-1)\nu_i, t) \\
&\quad + Q(S + (j-1)\nu_i, S + (j-1)\nu_i)
\end{aligned}\right\} \quad (25.17)$$

and

$$\left.\begin{aligned}
&- Q(S + (j-1)\nu_i, t) + Q(S + (j-1)\nu_i, S + (j-1)\nu_i) \\
&= - Q(S+(2j-2)\nu_{i+1}, t) + Q(S+(2j-2)\nu_{i+1}, S+(2j-2)\nu_{i+1})
\end{aligned}\right\} \quad (25.18)$$

since $(j-1)\nu_i = (2j-2)\nu_{i+1}$. (25.16), (25.17), (25.3) imply that

$$\left.\begin{aligned}
&\|L_{i,j}(t)\| \\
&= \|Q(S+(2j-2)\nu_{i+1}, S+(2j-1)\nu_{i+1}) - Q(S+(2j-2)\nu_{i+1}, t) \\
&\quad - Q(S+(2j-1)\nu_{i+1}, S+(2j-1)\nu_{i+1}) + Q(S+(2j-1)\nu_{i+1}, t)\| \\
&\leq \psi_1(\nu_{i+1})\,\psi_2(\nu_{i+1}) \\
&\quad \text{for} \ \ S+(2j-1)\nu_{i+1} \leq t \leq S+2j\nu_{i+1}, i \in \mathbb{N}_0, j = 1, 2, \ldots, 2^i.
\end{aligned}\right\} \quad (25.19)$$

Further (cf. (25.5)),

$$\|L_{i,j}(t)\| = \|L_{i,j}(S + 2j\nu_{i+1})\| \quad \text{for} \ \ S + 2j\nu_{i+1} \leq t \leq c. \quad (25.20)$$

(25.17)–(25.20) imply that

$$\|L_{i,j}(t)\| \leq \psi_1(\nu_{i+1})\,\psi_2(\nu_{i+1}) \text{ for } t \in [a, c], \ i \in \mathbb{N}_0, \ j = 1, 2, \ldots, 2^i. \quad (25.21)$$

By (25.10), (25.14), (25.19)

$$\left.\begin{aligned} \|R_{i+1}(S,t,T) - R_i(S,t,T)\| &\le 2^i\,\psi_1(\nu_{i+1})\,\psi_2(\nu_{i+1}) \\ &\qquad \text{for } i\in\mathbb{N}_0,\ t\in[a,c], \end{aligned}\right\} \quad (25.22)$$

$$\left.\begin{aligned} \|R_{i+1}(S,t,T) - R_0(S,t,T)\| &\le \sum_{j=0}^{i} 2^j\,\psi_1(\nu_{j+1})\,\psi_2(\nu_{j+1}) \\ &\le \tfrac{1}{2}\,\Psi(T-S) \quad \text{for } i\in\mathbb{N}_0,\ t\in[a,c]. \end{aligned}\right\} \quad (25.23)$$

If $S \le t \le T$ then (cf. (25.5)–(25.7), (25.23))

$$R_0(S,t,T) = Q(S,t) - Q(S,S), \qquad (25.24)$$

$$\|R_{i+1}(S,t,T) - Q(S,t) + Q(S,S)\| \le \frac{1}{2}\,\Psi(T-S). \qquad (25.25)$$

By (25.8), (25.5) $P_i(a) = 0$ for $i\in\mathbb{N}_0$. By (25.10), (25.22)

$$\|P_{i+1}(t) - P_i(t)\| \le 2^i\,\psi_1(\mu_{i+1})\,\psi_2(\mu_{i+1}) \quad \text{for } i\in\mathbb{N}_0,\ t\in[a,c], \quad (25.26)$$

which implies that

$$\text{the limit in (25.11) exists} \qquad (25.27)$$

since the series $\displaystyle\sum_{i=0}^{\infty} 2^i\,\psi_1(\mu_{i+1})\,\psi_2(\mu_{i+1})$ is convergent. Let $i,\ell,m\in\mathbb{N}_0$, $m \le 2^i,\ i \ge \ell$,

$$S = a + m\,\mu_i, \quad T = S + (b-a)\,2^{-\ell}, \quad S \le t \le T. \qquad (25.28)$$

Observe that $\nu_{i-\ell} = \mu_i$. Then (cf. (25.7))

$$\left. \begin{aligned}
R_{i-\ell}(S,t,T) &= \sum_{j=1}^{2^{i-\ell}} \mathbf{Q}(S+(j-1)\,\nu_{i-\ell}, t, S+j\,\nu_{i-\ell}) \\
&= \sum_{j=1}^{2^{i-\ell}} \mathbf{Q}(a+(m+j-1)\,\mu_i, t, a+(m+j)\,\mu_i) \\
&= \sum_{j=m+1}^{m+2^{i-\ell}} \mathbf{Q}(a+(j-1)\,\mu_i, t, a+j\,\mu_i) \\
&= \sum_{j=1}^{m+2^{i-\ell}} \mathbf{Q}(a+(j-1)\,\mu_i, t, a+j\,\mu_i) \\
&\quad - \sum_{j=1}^{m} \mathbf{Q}(a+(j-1)\,\mu_i, t, a+j\,\mu_i).
\end{aligned} \right\} \quad (25.29)$$

By (25.8)

$$\left. \begin{aligned}
P_i(t) &= \sum_{j=1}^{2^i} \mathbf{Q}(a+(j-1)\,\mu_i, t, a+j\,\mu_i) \\
&= \sum_{j=1}^{m+2^{i-\ell}} \mathbf{Q}(a+(j-1)\,\mu_i, t, a+j\,\mu_i)
\end{aligned} \right\} \quad (25.30)$$

since $t \leq T = a + (m+2^{i-\ell})\,\mu_i$ which implies that

$$\mathbf{Q}(a+(j-1)\,\mu_i, t, a+j\,\mu_i) = 0 \quad \text{if} \quad a+(j-1)\,\mu_i \geq T, \quad \text{i.e. if} \quad j \geq m+1+2^{i-\ell}.$$

Moreover,

$$\sum_{j=1}^{m} \mathbf{Q}(a+(j-1)\,\mu_i, t, a+j\,\mu_i) = \sum_{j=1}^{m} \mathbf{Q}(a+(j-1)\,\mu_i, S, a+j\,\mu_i)$$

since $t \geq S$ and

$$\begin{aligned}
\mathbf{Q}(a+(j-1)\,\mu_i, t, a+j\,\mu_i) &= \mathbf{Q}(a+(j-1)\,\mu_i, a+j\,\mu_i, a+j\,\mu_i) \\
&= \mathbf{Q}(a+(j-1)\,\mu_i, S, a+j\,\mu_i) \quad \text{for} \quad a+j\,\mu_i \leq S, \quad \text{i.e. for} \quad j \leq m.
\end{aligned}$$

Therefore

$$\sum_{j=1}^{m} \mathbf{Q}(a+(j-1)\,\mu_i, t, a+j\,\mu_i) = P_i(S). \quad (25.31)$$

(25.29)–(25.31) imply that

$$R_{i-\ell}(S,t,T) = P_i(t) - P_i(S).\tag{25.32}$$

since $\mathbf{Q}(a + (j-1)\,\mu_i, t, a + j\,\mu_i) = 0$ if $a + j\,\mu_i > T$, i.e. $j > m + 2^{i-\ell}$. Now, (25.32) and (25.23) give

$$\left.\begin{array}{r} \|P_i(t) - P_i(S) - Q(S,t) + Q(S,S)\| \le \tfrac{1}{2}\,\Psi(T-S) \\[4pt] \text{for } S \le t \le T, \quad S,T \text{ fulfilling (25.28).} \end{array}\right\}\tag{25.33}$$

If $k \in \mathbb{N}_0$, $S = a + m\,\mu_i$ then $S = a + 2^k\,m\,\mu_{i+k}$. Hence

$$\|P_{i+k}(t) - P_{i+k}(S) - Q(S,t) + Q(S,S)\| \le \frac{1}{2}\,\Psi(T-S).$$

The limit procedure for $k \to \infty$ (cf. also (25.27)) implies that

$$\|P(t) - P(S) - Q(S,t) + Q(S,S)\| \le \frac{1}{2}\,\Psi(T-S)\tag{25.34}$$

if S,t,T fulfil (25.28) for suitable i,ℓ,m. But P and Q are continuous and (25.34) is valid if $S,T \in [a,b]$, $S \le T$, $t = T$. Hence

$$\left.\begin{array}{r} \|P(t) - P(S) - Q(S,t) + Q(S,S)\| \le \tfrac{1}{2}\,\Psi(t-S) \\[4pt] \text{for } S,t \in [a,b], \; S \le t. \end{array}\right\}\tag{25.35}$$

If $a \le \bar{t} \le S \le b$ then by (25.34)

$$\|P(S) - P(\bar{t}) - Q(\bar{t},S) + Q(\bar{t},\bar{t})\| \le \frac{1}{2}\,\Psi(S-\bar{t}).\tag{25.36}$$

By (25.3)

$$\| - Q(\bar{t},S) + Q(\bar{t},\bar{t}) + Q(S,S) - Q(S,\bar{t})\| \le \psi_1(S-\bar{t})\,\psi_2(S-\bar{t}),$$

which together with (25.36) implies that

$$\left.\begin{array}{l} \|P(\bar{t}) - P(S) - Q(S,\bar{t}) + Q(S,S)\| \\[4pt] \qquad \le \tfrac{1}{2}\,\Psi(S-\bar{t}) + \psi_1(S-\bar{t})\,\psi_2(S-\bar{t}). \end{array}\right\}\tag{25.37}$$

Moreover (cf. (7.15), (7.19)),

$$\frac{1}{\sigma}\,\Psi(\sigma) \to 0, \quad \frac{1}{\sigma}\,\psi_1(\sigma)\,\psi_2(\sigma) \to 0 \quad \text{for } \sigma \to 0+.$$

Hence (cf. (25.35), (25.37)) for $\varepsilon > 0$ there exists $\xi > 0$ such that

$$\|P(t) - P(S) - Q(t, S) + Q(S, S)\| \leq \varepsilon (t - S)$$
$$\text{for } a \leq S < t \leq \min\{S + \xi, b\}$$

and

$$\|P(\bar{t}) - P(S) - Q(\bar{t}, S) + Q(S, S)\| \leq \varepsilon (S - \bar{t})$$
$$\text{for } a \leq \bar{t} < S \leq \min\{\bar{t} + \xi, b\}.$$

Therefore (25.12) is correct. (25.13) is a consequence of (25.12). The proof is complete. $\square$

Chapter 26

On equality of two integrals

26.1. Notation. Let

$$[a,b] \subset \mathbb{R}, \ \widehat{R} > 0, \ U: [a,b]^2 \to X, \ \bar{U}: [a,b]^2 \to X, \ \widehat{\Psi}: \mathbb{R}_0^+ \to \mathbb{R}_0^+.$$

26.2. Lemma. *Assume that*

U *is* SR-*integrable on* $[a,b]$, \hfill (26.1)

$$\frac{\widehat{\Psi}(\sigma)}{\sigma} \to 0 \quad for \ \sigma \to 0+, \tag{26.2}$$

$$\|U(\tau,t) - U(\tau,\bar{t}) - \bar{U}(\tau,t) + \bar{U}(\tau,\bar{t})\| \le \widehat{\Psi}(t - \bar{t}) \ for \ a \le \bar{t} \le \tau \le t \le b. \tag{26.3}$$

Then

$\bar{U}$ *is* SR-*integrable on* $[a,b]$, \hfill (26.4)

$$(\text{SR}) \int_a^T D_t U(\tau,t) = (\text{SR}) \int_a^T D_t \bar{U}(\tau,t) \quad for \ T \in [a,b]. \tag{26.5}$$

Proof. Let $u: [a,b] \to X$ be an SR-primitive of U. Let $a < T \le b$, $\varepsilon > 0$. There exists $\xi_1 > 0$ such that

$$\widehat{\Psi}(\sigma) \le \varepsilon \sigma \quad for \ 0 < \sigma \le \xi_1.$$

Moreover, there exists ξ_2, $0 < \xi_2 \le \xi_1$, such that

$$\sum_{\mathcal{S}} \|u(s) - u(\bar{s}) - U(\sigma,s) + U(\sigma,\bar{s})\| \le \varepsilon$$

for an arbitrary ξ_2-fine partition $\mathcal{S} = \{([\bar{s},s],\sigma)\}$ of $[a,b]$. Let $\mathcal{R} = \{([\bar{r},r],\rho)\}$

185

be a ξ_2-fine partition of $[a, b]$. Then

$$\sum_{\mathcal{R}} \|u(r) - u(\bar{r}) - \overline{U}(\rho, r) + \overline{U}(\rho, \bar{r})\|$$

$$\leq \sum_{\mathcal{R}} \|u(r) - u(\bar{r}) - U(\rho, r) + U(\rho, \bar{r})\|$$

$$+ \sum_{\mathcal{R}} \|U(\rho, r) - U(\rho, \bar{r}) - \overline{U}(\rho, r) + \overline{U}(\rho, \bar{r})\|$$

$$\leq \varepsilon + \sum_{\mathcal{R}} \widehat{\Psi}(r - \bar{r}) \leq \varepsilon (1 + b - a)$$

i.e. $\overline{U}$ is SR-integrable and (26.5) holds. $\square$

Chapter 27

A class of generalized ordinary differential equations with a restricted right hand side

27.1. Definition. The right-hand side of the GODE

$$\frac{\mathrm{d}}{\mathrm{d}\,t} = \mathrm{D}_t F(x,\tau,t)$$

is called restricted if F is independent of τ.

27.2. Remark. GODEs of this type were met in Theorem 23.11. In this chapter an analogous result will be proved for the class of GODEs from Chapters 8–13.

27.3. Notation. Let $[a,b] \subset \mathbb{R}$, $R > 0$, $\widehat{R} = 8R$. Let ψ_1, ψ_2, Ψ be functions which were introduced in Chapter 7 and let $G : B(8R) \times [a,b]^2 \to X$ fulfil (8.2)–(8.7). Put

$$\widehat{\Psi}(\sigma) = \frac{1}{2}\,\Psi(\sigma) + 2\,\psi_1(\sigma)\,\psi_2(\sigma) \quad \text{for} \ \ \sigma \in \mathbb{R}_0^+ , \tag{27.1}$$

$$G^\circ(x,T) = (\mathrm{SR}) \int_a^T \mathrm{D}_t G(x,\tau,t) \quad \text{for} \ \ x \in B(8R),\ T \in [a,b]. \tag{27.2}$$

27.4. Lemma. $G(x,.,.)$ *is* SR-*integrable* (x *being kept fixed*). *Moreover,*

$$\left.\begin{array}{c} \|G^\circ(x,T) - G^\circ(x,S) - G(x,S,T) + G(x,S,S)\| \\[4pt] \leq \widehat{\Psi}(|T{-}S|) + \psi_1(|T{-}S|)\,\psi_2(|T{-}S|) \\[4pt] \text{for} \ \ x \in B(8R),\ S,T \in [a,b]. \end{array}\right\} \tag{27.3}$$

Proof. The SR-integrability of $G(x,.,.)$ and (27.3) are consequences of Theorem 25.2 where $Q(S,T) = G(x,S,T)$, $P(T) = G^\circ(x,T)$ for $S,T \in [a,b]$. $\qquad\square$

27.5. Theorem.

(i) If $u\colon [a,b] \to B(\widehat{R})$ is an SR-*solution of the GODE*

$$\frac{\mathrm{d}}{\mathrm{d}t} x = \mathrm{D}_t G(x, \tau, t) \qquad (27.4)$$

then it is an SR-*solution of*

$$\frac{\mathrm{d}}{\mathrm{d}t} x = \mathrm{D}_t G^\circ(x, t). \qquad (27.5)$$

(ii) If $v\colon [a,b] \to B(\widehat{R})$ is an SR-*solution of the GODE* (27.5) *then it is an* SR-*solution of* (27.4) *as well.*

Proof.   (i) is a consequence of the definition of an SR-solution of a GODE and of Lemmas 26.2, 27.4 (where $U(\tau, t) = G(u(\tau), \tau, t)$, $\overline{U} = G^\circ(u(\tau), t)$) since $\frac{1}{\sigma}\Psi(\sigma) \to 0$, $\frac{1}{\sigma}\psi_1(\sigma)\,\psi_2(\sigma) \to 0$ for $\sigma \to 0+$ (cf. (7.17), (7.19)). The proof of (ii) is analogous. $\square$

Appendix A

Some elementary results

A.1. Lemma (Gronwall). *Let* $S > 0$, $\eta > 0$, $\omega : [0, S] \to \mathbb{R}_0^+$, $\xi : [0, S] \to \mathbb{R}_0^+$, ω, ξ *being continuous. Assume that*

$$\xi(t) \leq \eta + \int_0^t \omega(s)\,\xi(s)\,\mathrm{d}s \quad \text{if } 0 \leq t \leq S. \tag{A.1}$$

Then

$$\xi(t) \leq \eta \exp\left(\int_0^t \omega(s)\,\mathrm{d}s\right) \quad \text{if } 0 \leq t \leq S. \tag{A.2}$$

Comment. (A.2) is proved by integration of

$$\frac{\omega(t)\,\xi(t)}{\eta + \int_0^t \omega(s)\,\xi(s)\,\mathrm{d}s} \leq \omega(t) \quad \text{if } 0 \leq t \leq S.$$

Lemma A.1 is valid if ω is Lebesgue integrable and if ξ is bounded, measurable.

A.2. Lemma. *Let* $\lambda : \mathbb{R} \to X$. *Assume that*

$$\left. \begin{array}{c} \lambda \text{ is continuous,} \\[2mm] \|\lambda(t)\| \leq 1, \quad \lambda(t+1) = \lambda(t) \text{ for } t \in \mathbb{R}, \quad \int_0^1 \lambda(\tau)\,\mathrm{d}\tau = 0. \end{array} \right\} \tag{A.3}$$

Then there exists $\Lambda : \mathbb{R} \to X$ *such that*

$$\frac{\mathrm{d}\Lambda}{\mathrm{d}t}(t) = \lambda(t), \quad \Lambda(t+1) = \Lambda(t), \quad \int_t^{t+1} \Lambda(\tau)\,\mathrm{d}\tau = 0 \text{ for } t \in \mathbb{R} \tag{A.4}$$

and

$$\|\Lambda(t)\| \leq \tfrac{1}{4}, \quad t \in \mathbb{R}. \tag{A.5}$$

Proof. $\displaystyle\int_t^{t+1} \lambda(\tau)\,\mathrm{d}\tau = \int_0^1 \lambda(\tau)\,\mathrm{d}\tau = 0$ for $t \in \mathbb{R}$ by (A.3). Put

$$\bar\Lambda(t) = \int_0^t \lambda(\tau)\,\mathrm{d}\tau, \quad \Lambda(t) = \bar\Lambda(t) - \int_0^1 \bar\Lambda(t)\,\mathrm{d}t\,.$$

Then (A.4) is fulfilled. Moreover,

$$\bar\Lambda(t) - \bar\Lambda(s) = \int_s^t \lambda(\sigma)\,\mathrm{d}\sigma, \quad \bar\Lambda(t+1) = \bar\Lambda(t) \ \text{ for } t \in \mathbb{R}\,,$$

$$\|\bar\Lambda(t)) - \bar\Lambda(s)\| \le |t - s| \quad \text{for } s, t \in \mathbb{R}\,,$$

$$\Lambda(t) = \bar\Lambda(t) - \int_0^1 \bar\Lambda(\sigma)\,\mathrm{d}\sigma\,,$$

$$= \bar\Lambda(t)) - \int_{t-\frac{1}{2}}^{t+\frac{1}{2}} \bar\Lambda(\sigma)\,\mathrm{d}\sigma = \int_{t-\frac{1}{2}}^{t+\frac{1}{2}} [\bar\Lambda(t)) - \bar\Lambda(\sigma)]\,\mathrm{d}\sigma\,.$$

Hence

$$\|\Lambda(t)\| \le \int_{t-\frac{1}{2}}^{t+\frac{1}{2}} |t - \sigma|\,\mathrm{d}\sigma = \tfrac{1}{4} \quad \text{for } t \in \mathbb{R}$$

and (A.5) holds. $\qquad\qquad\square$

Appendix B

Trifles from functional analysis

B.1. Lemma. *Let*

$$0 < R_1 < R_2, \quad Q : B(R_2) \to X, \tag{B.1}$$

and assume that

$$\left. \begin{array}{l} \|x\,Q\| \leq \tfrac{2}{3}\,(R_2 - R_1) \quad \text{for } x \in B(R_1)\,, \\[2mm] \|x\,Q - \hat{x}\,Q\| \leq \tfrac{1}{3}\|x - \hat{x}\| \quad \text{for } x, \hat{x} \in B(R_2)\,, \end{array} \right\} \tag{B.2}$$

$x\,Q$ being the value of Q at x.

Then for every $z \in B(R_1)$ there exists $x \in B(R_2)$ such that

$$x + x\,Q = z. \tag{B.3}$$

Proof. Let $z \in B(R_1)$. Put

$$x_0 = z \tag{B.4}$$

and if $\ell \in \mathbb{N}_0$, $x_\ell \in B(R_2)$, put

$$x_{\ell+1} = z - x_\ell\,Q. \tag{B.5}$$

Then (cf. (B.2))

$$x_1 - x_0 = -z\,Q\,, \tag{B.6}$$

$$\|x_1 - x_0\| \leq \frac{2}{3}\,(R_2 - R_1), \quad \|x_{\ell+1} - x_\ell\| \leq \frac{2}{3}\,(R_2 - R_1)\,(\frac{1}{3})^\ell,$$

$$\|x_m - x_\ell\| \leq \frac{2}{3}\,(R_2 - R_1)\,(\frac{1}{3})^\ell\,(1 + \frac{1}{3} + (\frac{1}{3})^2 + \cdots)$$

$$\leq (\frac{1}{3})^\ell\,(R_2 - R_1) \quad \text{for } m > \ell,$$

$$\|x_m - x_0\| \le R_2 - R_1\,,$$

$x_m \in B(R_2)$ for $m \in \mathbb{N}_0$ and $(x_m,\ m = 0, 1, 2, \dots)$ is a Cauchy sequence. Define

$$x = \lim_{m \to \infty} x_m\,.$$

Then x fulfils (B.3) by the limit procedure in (B.5).   □

B.2. Remark. Let $[a, b] \subset \mathbb{R}$, $u : [a, b] \to X$. u is called a *function of bounded variation* on $[a, b]$ if there exists $\varkappa \in \mathbb{R}^+$ such that

$$\sum_{i=1}^{k} \|u(t_{i+1}) - u(t_i)\| \le \varkappa$$

for any $k \in \mathbb{N}$ and any sequence $a = t_1 < t_1 < \dots < t_k \le b$. The infimum of such $\varkappa$ is called the *variation* of u on $[a, b]$ and denoted by $\mathrm{var}\,(u, [a, b])$ or $\mathrm{var}\,u$. If u is a function of bounded variation on $[a, b]$ then $u(\tau+) = \lim\limits_{t \to \tau, t > \tau} u(t)$ exists for $a \le \tau < b$ and $u(\tau-) = \lim\limits_{t \to \tau, t < \tau} u(t)$ exists for $a < \tau \le b$. Let $\varepsilon > 0$. The set of τ such that either $\|u(\tau+) - u(\tau)\| \ge \varepsilon$ or $\|u(\tau) - u(\tau-)\| \ge \varepsilon$ is finite.

B.3. Remark. Let $[a, b] \subset \mathbb{R}$, $v : [a, b] \to X$. v is called a *regulated function* on $[a, b]$ if

$$v(\tau+) = \lim_{t \to \tau, t > \tau} v(t) \quad \text{exists for } a \le \tau < b$$

and

$$v(\tau-) = \lim_{t \to \tau, t < \tau} v(t) \quad \text{exists for } a < \tau \le b\,.$$

Let v be a regulated function. Then v is bounded and for every $\varepsilon > 0$ the number of τ such that either $\|v(\tau+) - v(\tau)\| \ge \varepsilon$ or $\|v(\tau) - v(\tau-)\| \ge \varepsilon$ is finite.

Bibliography

[Bartle, Sherbert (2000)]
    Bartle, R. G. and Sherbert, D. R., *Introduction to Real Analysis*. Wiley, New York, 2000.

[Gordon (1994)]
    Gordon, R. A., *The integrals of Lebesgue, Denjoy, Perron, and Henstock*. American Mathematical Society, 1994.

[Henstock (1988)]
    Henstock, R., *Lectures on the Theory of Integration*. World Scientific, Singapore, 1988.

[Henstock (1991)]
    Henstock, R., *The General Theory of Integration*. Clarendon Press, Oxford, 1991.

[Jarník (1961a)]
    Jarník, J., *On some assumptions of the theorem on the continuous dependence on a parameter*. Čas. Pěst. Mat. 86 (1961), 404–414.

[Jarník (1961b)]
    Jarník, J., *On a certain modification of the theorem on the continuous dependence on a parameter*. Čas. Pěst. Mat. 86 (1961), 415–424.

[Jarník (1965)]
    Jarník, J., *Dependence of solutions of a class of differential equations of the second order on a parameter*. Czechoslovak Math. J., 15 (90), (1965), 124–160.

[Kapitza (1951a)]
    Kapitza, P. L., *Pendulum with vibrating suspension* (in Russian). Uspechi fiz. nauk, XLIV, 1 (1951), 7–20.

[Kapitza (1951b)]
    Kapitza, P. L., *Dynamic stability of a pendulum with an oscilatting point of suspension* (in Russian). Zhurnal eksperimental'noi i teoreticheskoi fiziki, 21, 5 (1951), 588–597. Translation in: *Collected Papers by P. L. Kapitza*, vol. 2, 714–726, Pergamon Press, London, 1965.

[Kurzweil (1957)]
Kurzweil, J., *Generalized ordinary differential equations and continuous dependence on a parameter.* Czechoslovak Math. J., 7 (82),(1957), 418–449.

[Kurzweil (1958)]
Kurzweil, J., *On integration by parts.* Czechoslovak Math. J.,8 (82),(1958), 356–359.

[Kurzweil (1980)]
Kurzweil, J., *Nichtabsolut konvergente Integrale.* Teubner-Verlag, Leipzig, 1980.

[Kurzweil (2000)]
Kurzweil, J., *Henstock-Kurzweil Integration : its relation to topological vector spaces.* World Scientific Publishing Co. Inc., River Edge, NJ, 2000.

[Lee P.Y. (1989)]
Lee, P. Y., *Lanzhou Lectures on Henstock Integration.* World Scientific, Singapore, 1989.

[Lee P.Y., Výborný (2000)]
Lee, P. Y. and Výborný, R., *The Integral : An Easy Approach after Kurzweil and Henstock.* Cambridge Univ. Press, Cambridge, 2000.

[Lojasiewicz (1955)]
Lojasiewicz, S., *Sur un effet asymptotique dans les équations différentielles dont les seconds membres contiennent des termes périodiques de pulsation et d'amplitude tendant a l'infini.* Ann. Pol. Math 1 (1955), 388–413.

[Mitropolskij (1971)]
Mitropolskij, Yu. A, *Averaging in Nonlinear Mechanics* (in Russian). Naukova Dumka, Kiev 1971.

[Monteiro, Tvrdý (2011)]
Monteiro, G. and Tvrdý, M. *On Kurzweil-Stieltjes integral in Banach space.* Math. Bohem., to appear.

[Pfeffer (1993)]
Pfeffer, W. F., *The Riemann Approach to Integration : Local geometric theory.* Cambridge University Press, 1993.

[Schwabik (1992)]
Schwabik, Š., *Generalized Ordinary Differential Equations.* World Scientific, Singapore, 1992.

[Schwabik (2001)]
Schwabik, Š., *A note on integration by parts for abstract Perron-Stieltjes integrals.* Math. Bohem. **126** 2001, 613–629.

[Schwabik, Ye (2005)]
Schwabik, Š. and Ye, G., *Topics in Banach space Integration.* World Scientific, Singapore, 2005.

Symbols

$D_1 h(x,t)$, 12
$D_2 h(x,t)$, 12
$D^2 h(x)$, 12
$D h(x)$, 12

$[a,b]$, 12
$\mathcal{A}(\sigma)$, 43

$B(r)$, 12, 37
$\mathcal{B}(\sigma)$, 43

$\Delta_v F(x,p)$, 47
$\Delta_v f(x)$, 47
dist (G, G^*), 48
$D\, p \circ q(x)$, 16

$\mathcal{E}_{S,\sigma}$, 69

$\Phi(\sigma)$, 43
$\varphi(\sigma)$, 43

G, 47, 51
$\mathbf{G}$, 51
$\widetilde{G}$, 51

$\mathbb{N}$, 12
$\mathbb{N}_0$, 12

Ω, 48, 79

$[p,q](x)$, 16
$\Psi(\sigma)$, 43
ψ_1, ψ_2, 43

$\mathbb{R}$, 12
$\mathbb{R}^+$, 12
$\mathbb{R}_0^+$, 12

$U(\sigma, \cdot)$, 27
$\widehat{U}$, 133

W, 69

$\|x\|$, 27
X, 27
$x\, V_i(S,t,T)$, 53
$x Y_i(S,t,T)$, 54

Subject index